RESEARCH NEEDS IN SUBSURFACE SCIENCE

U.S. Department of Energy's
Environmental Management Science Program

Board on Radioactive Waste Management
Water Science and Technology Board

National Research Council

NATIONAL ACADEMY PRESS
Washington, D.C.

NOTICE: The project that is the subject of this report was approved by the Governing Board of the National Research Council, whose members are drawn from the councils of the National Academy of Sciences, the National Academy of Engineering, and the Institute of Medicine. The members of the committee responsible for the report were chosen for their special competences and with regard for appropriate balance.

This study was supported by Contract/Grant No DE-FC01-94EW54069/R between the National Academy of Sciences and The U.S. Department of Energy. Any opinions, findings, conclusions, or recommendations expressed in this publication are those of the author(s) and do not necessarily reflect the views of the organizations or agencies that provided support for the project.

International Standard Book Number 0-309-06646-8

Additional copies of this report are available from National Academy Press, 2101 Constitution Avenue, N.W., Lockbox 285, Washington, D.C. 20055; (800) 624-6242 or (202) 334-3313 (in the Washington metropolitan area); Internet: http://www.nap.edu

COVER IMAGE: Mercury contamination in soil at the Y-12 plant at the Oak Ridge Reservation. The mercury is visible as small droplets in the dark layer near the center of the photograph. SOURCE: Oak Ridge Reservation.

Printed in the United States of America

THE NATIONAL ACADEMIES

National Academy of Sciences
National Academy of Engineering
Institute of Medicine
National Research Council

The **National Academy of Sciences** is a private, nonprofit, self-perpetuating society of distinguished scholars engaged in scientific and engineering research, dedicated to the furtherance of science and technology and to their use for the general welfare. Upon the authority of the charter granted to it by the Congress in 1863, the Academy has a mandate that requires it to advise the federal government on scientific and technical matters. Dr. Bruce M. Alberts is president of the National Academy of Sciences.

The **National Academy of Engineering** was established in 1964, under the charter of the National Academy of Sciences, as a parallel organization of outstanding engineers. It is autonomous in its administration and in the selection of its members, sharing with the National Academy of Sciences the responsibility for advising the federal government. The National Academy of Engineering also sponsors engineering programs aimed at meeting national needs, encourages education and research, and recognizes the superior achievements of engineers. Dr. William A. Wulf is president of the National Academy of Engineering.

The **Institute of Medicine** was established in 1970 by the National Academy of Sciences to secure the services of eminent members of appropriate professions in the examination of policy matters pertaining to the health of the public. The Institute acts under the responsibility given to the National Academy of Sciences by its congressional charter to be an adviser to the federal government and, upon its own initiative, to identify issues of medical care, research, and education. Dr. Kenneth I. Shine is president of the Institute of Medicine.

The **National Research Council** was organized by the National Academy of Sciences in 1916 to associate the broad community of science and technology with the Academy's purposes of furthering knowledge and advising the federal government. Functioning in accordance with general policies determined by the Academy, the Council has become the principal operating agency of both the National Academy of Sciences and the National Academy of Engineering in providing services to the government, the public, and the scientific and engineering communities. The Council is administered jointly by both Academies and the Institute of Medicine. Dr. Bruce M. Alberts and Dr. William A. Wulf are chairman and vice chairman, respectively, of the National Research Council.

COMMITTEE ON SUBSURFACE CONTAMINATION AT DOE COMPLEX SITES

JANE C. S. LONG, *Chair,* University of Nevada, Reno
JAMES K. MITCHELL, *Vice-Chair,* Virginia Polytechnic Institute and State University, Blacksburg
RANDALL J. CHARBENEAU, University of Texas, Austin
JEFFREY J. DANIELS, Ohio State University, Columbus
JOHN N. FISCHER, Hydrologic Consultant, Oakton, Virginia
TISSA H. ILLANGASEKARE, Colorado School of Mines, Golden
AARON L. MILLS, University of Virginia, Charlottesville
DONALD T. REED, Argonne National Laboratory, Chicago, Illinois
JEROME SACKS, National Institute of Statistical Sciences, Research Triangle Park, North Carolina
BRIDGET R. SCANLON, University of Texas, Austin
LEON T. SILVER, California Institute of Technology, Pasadena
CLAIRE WELTY, Drexel University, Philadelphia, Pennsylvania

STAFF

KEVIN D. CROWLEY, Study Director
STEPHEN D. PARKER, Director, Water Science and Technology Board
SUSAN B. MOCKLER, Research Associate
PATRICIA A. JONES, Senior Project Assistant

BOARD ON RADIOACTIVE WASTE MANAGEMENT

STAFF

Preface

The development of this report has provided an opportunity for committee members to examine and obtain an overview of a major national environmental issue—subsurface contamination in the DOE complex. The committee faced a daunting task in making recommendations to the Environmental Management Science Program about future research emphases to address DOE's subsurface contamination problems. To do this, we needed to obtain an overview of the problems and a detailed understanding of the major clean-up issues. In addition, we needed to understand how the Environmental Management Science Program had developed so far, whether it related well to the problems as we understood them, and its relationship to environmental remediation research done elsewhere. Finally, we were to complete this task in approximately one year with a limited number of site visits.

Clearly, we could never have accomplished this task without the complete cooperation of the DOE and National Laboratory staff. We owe major thanks to a large number of persons (see Appendix B) who prepared presentations and organized visits that informed our process. A great deal of effort was spent to support us, and I would like to thank all of these people for their frankness and insights. I would especially like to recognize the efforts of Mark Gilbertson and Roland Hirsch from DOE headquarters; Roy Gephart, John Zacara, and Karl Fecht from Hanford; Tom Williams from the Idaho National Engineering and Environmental Laboratory; and Tom Hicks and Tom Temples from Savannah River for their support of the committee.

I have served on a number of excellent National Research Council committees, but I found the support provided by committee staff on this study was beyond any level of service I have ever experienced. Study director Kevin Crowley made this difficult task possible. Without his understanding, sense of group dynamics, and very significant level of effort there would have been no possibility of finishing this report. We were also provided excellent research and logistical support by the staff of the Board on Radioactive Waste Management and Water Science and Technology Board, most notably Steve Parker, Patricia Jones, and Susan Mockler.

We were greatly privileged to have Jim Mitchell serve as the committee's vice-chair. Jim was the conscience of the committee and played a critical role in keeping us on course throughout our deliberations. His careful analysis, insight, and review provided quality to our product. It was a great treat to work with Jim.

Normal committee dynamics are such that a few people do a disproportionate share of the work. This committee was an exception to that rule; the members all contributed and all did the assignments we gave them. The committee was unusually productive and creative, and its members contributed not only their knowledge and understanding, but they also listened to others and incorporated this information into a consensus. I learned a great deal from my committee colleagues, and my sense is that the entire committee found the process beneficial.

The committee's review left some very clear impressions concerning the scope of DOE's subsurface contamination problems. As noted in Chapter 2 of this report, the committee concluded that much of the contamination that is now in the subsurface at major DOE sites will not be removed by any active remediation efforts. The huge scale of the "environmental insult" (to quote committee member Lee Silver) and the extraction of contamination on the scales required would require a major decrease in entropy and would simply not be possible. This means that a major focus of coming to terms with the problem has to be understanding, predicting, and containing the subsurface contamination. These issues are paramount in site closure. They have received insufficient attention from the EMSP in the past and are a major focus of this report.

Secondly, the committee recognized that the amount of contamination that is contained in surface and near-surface facilities at DOE sites is massive compared to that which has already leaked into the subsurface. Millions of gallons of waste and millions of curies of radioactivity are currently in storage at DOE sites and, if this waste is not managed correctly, it could potentially become a major source of future subsurface contamination. Clearly, an important lesson DOE can learn from its current subsurface contamination problems is to not repeat the mistakes of the past. It is true that DOE no longer places high-level nuclear waste in barrels that are dumped into topographic lows (see Sidebar 2.5), but DOE is placing new land disposal facilities in regions that have generated massive contaminant plumes in the past (see, for example, Sidebar 2.9). During the course of this study, the committee saw no institutional process to address the question, "How should the results and impacts of what was done in the past inform the decisions of the future?" The committee recognizes that DOE cannot change what was done in the past. DOE can, however, make better decisions in the future. The committee believes that a very important role for research sponsored by the Environmental Management Science Program is to provide the information DOE will need to make technically sound and responsible waste management decisions in the future.

Jane C. S. Long, Chair

List of Report Reviewers

This report has been reviewed in draft form by individuals chosen for their diverse perspectives and technical expertise, in accordance with procedures approved by the NRC's Report Review Committee. The purpose of this independent review is to provide candid and critical comments that will assist the institution in making the published report as sound as possible and to ensure that the report meets institutional standards for objectivity, evidence, and responsiveness to the study charge. The review comments and draft manuscript remain confidential to protect the integrity of the deliberative process. We wish to thank the following individuals for their participation in the review of this report:

Susan Brantley, Pennsylvania State University
Helen Dawson, U.S. Environmental Protection Agency
John Fountain, State University of New York
Robert Huggett, Michigan State University
Philip Palmer, DuPont (retired)
Frank Schwartz, Ohio State University
John Taylor, Electric Power Research Institute (retired)
Peter Wierenga, University of Arizona

Although the reviewers listed above have provided many constructive comments and suggestions, they were not asked to endorse the conclusions or recommendations, nor did they see the final draft of the report before its release. The review of this report was overseen by George Hornberger, appointed by the Commission on Geosciences, Environment, and Resources, and Paul Barton, appointed by the Report Review Committee, who were responsible for making certain that an independent examination of this report was carried out in accordance with NRC procedures and that all review comments were carefully considered. Responsibility for the final content of this report rests entirely with the authoring committee and the NRC.

Contents

RESEARCH NEEDS IN SUBSURFACE SCIENCE

Summary

In the spring of 1998, the U.S. Department of Energy (DOE) requested that the National Academies convene a committee of experts to provide recommendations on the formulation of a long-term basic research program to address subsurface contamination problems at DOE sites (see Sidebar 1.1 in Chapter 1). In response to this request, a committee with expertise in basic research and research management was formed under the joint auspices of the National Research Council's Board on Radioactive Waste Management and Water Science and Technology Board. A summary of the committee's information-gathering activities and its conclusions and recommendations are presented in this report.

The report provides an overview of the subsurface contamination problems across the DOE complex and shows by examples from the six largest DOE sites (Hanford Site, Idaho Engineering and Environmental Laboratory, Nevada Test Site, Oak Ridge Reservation, Rocky Flats Environmental Technology Site, and Savannah River Site) how advances in scientific and engineering knowledge can improve the effectiveness of the cleanup effort (see Chapter 2). The committee analyzed the current Environmental Management (EM) Science Program portfolio of subsurface research projects (see Chapter 3) to assess the extent to which the program is focused on DOE's contamination problems. This analysis employs an organizing scheme that provides a direct linkage between basic research in the EM Science Program and applied technology development in DOE's Subsurface Contaminants Focus Area. The committee also reviewed related research programs in other DOE offices and other federal agencies (see Chapter 4) to determine the extent to which they are focused on DOE's subsurface contamination problems. On the basis of these analyses, the report singles out the highly significant subsurface contamination knowledge gaps and research needs that the EM Science Program must address if the DOE cleanup program is to succeed.

Significant amounts of subsurface contaminants are likely to remain even after DOE's cleanup program is completed.

Subsurface Contamination at DOE Sites

Nuclear weapons production has resulted in the contamination of the large DOE sites. This contamination exists today in a wide range of forms and locations—including contaminated waste burial grounds; contaminated soil, sediment, and rock; and contaminated groundwater—and is frequently difficult to locate, characterize, and remediate. Significant amounts of subsurface contaminants are likely to remain even after DOE's cleanup program is completed.

The committee concluded that subsurface contamination is an enormously difficult cleanup problem that represents a potentially large future mortgage for the nation. This mortgage could, however, be reduced significantly through the development and application of new and improved technologies. The development of such technologies will require advances in basic understanding of the complex natural systems at DOE sites and the nature of the contaminants there. Given the long-term nature of the cleanup mission and its projected cost—the program is planned to last until 2070 and cost on the order of $200 billion—the committee believes that DOE has sufficient time to do the basic research required to support the development and deployment of new cleanup technologies.

Given the long-term nature of the cleanup mission and its projected cost . . . the committee believes that DOE has sufficient time to do the basic research required to support the development and deployment of new cleanup technologies.

EM Science Program Research Portfolio

Since its establishment by Congress, the program has held four proposal competitions and has awarded about $225 million in funding, which puts it among the largest environmental research efforts in the federal government. The program has supported research projects relevant to many aspects of DOE's cleanup program, including subsurface contamination, high-level waste, and deactivation and decommissioning. The committee reviewed the research portfolio for fiscal years 1996 and 1997 and identified 91 projects that were relevant to DOE's subsurface contamination problems. The committee's review revealed some significant areas of strength. Fifty projects address organic contamination problems and 38 projects use a combination of field-, laboratory-, and modeling-based approaches. There appears to be a critical mass of projects covering remediation of subsurface contamination, especially treatment and destruction of organic contaminants through physical, chemical, and biological processes.

The most notable gaps in the current portfolio concern containment and validation.[1] These are two of the most significant problem areas in the DOE complex, because it is inevitable that DOE will have to manage much of its subsurface contamination in place. There also appear to be relatively few projects that address radionuclide and metal contamination problems.

Research Programs in Other Government Agencies

The committee gathered information on research programs in other DOE offices and other federal agencies to assess how they might contribute to solving DOE's subsurface contamination problems. The committee made the following observations in Chapter 4:

- The federal government is a major sponsor of basic research that is related either directly or indirectly to environmental problems. The committee identified almost 50 such programs in its survey (see Table 4.1).
- A large number and variety of programs across the federal government support research of direct relevance to the EM Science Program and DOE cleanup. The committee identified 18 such programs, many of which are focused on hazardous chemicals, especially volatile organic contaminants and non-aqueous phase liquids, and to a lesser extent on heavy metals. Many of these programs are also focused on remediation, especially bioremediation.
- With some notable exceptions, there appears to be significant overlap in scope among these 18 programs. It does not appear to the committee that these programs are being coordinated effectively among the agencies.

The committee concluded that there would be value-added to the EM Science Program and, ultimately, to DOE's cleanup efforts if there were better interagency coordination among these 18 research programs. The committee sees an opportunity for EM Science Program managers to promote and foster such coordination.

---

[1]The term "validation" is used to describe processes for testing a conceptual or predictive model to determine whether it adequately represents the system behavior of interest, and it is also applied to monitoring and testing to confirm the effectiveness of remediation actions. See Chapter 5.

Formulation of a Long-Term Research Program

The committee's recommendations for a long-term basic research program on subsurface contamination address the following issues:

- program vision,
- research emphases, and
- implementation.

The principal conclusions and recommendations are summarized below. Additional details can be found in Chapters 5 and 6.

Program Vision

The EM Science Program has been in existence for almost four years, but there does not appear to be a clear and agreed-upon vision for this program within DOE. If the program is to remain viable over the long term and to have a significant impact on the DOE cleanup mission, program managers must articulate a vision for the program that is supported both programmatically and financially by DOE upper management. The committee recommends that this vision include the following four elements:

1. The program objective should be to generate new knowledge to support DOE's mission to clean up its contaminated sites.
2. The program should be well connected to DOE's difficult cleanup problems.
3. A major focus of the program should continue to be on research to resolve DOE's subsurface contamination problems.
4. The program should have a long-term, multidisciplinary basic research[2] focus.

The committee defines "long term" as long enough to set ambitious goals to fill the knowledge gaps identified in Chapter 5 and to have reasonable expectations that those goals can be attained. In the committee's judgment, a time horizon on the order of a decade will be required to make cumulative progress on the knowledge gaps identified in Chapter 5, although shorter-term results of use to DOE's cleanup program will almost certainly be obtained over the lifetimes of individual research projects.

---

[2]Basic research creates new generic knowledge and is focused on long-term, rather than short-term, problems. See Sidebar 1.1 in Chapter 1.

Research Emphases

There are significant impediments to the successful completion of DOE's cleanup mission that can be removed through a focused, sustained, and adequately funded basic research program. Based on the analysis of DOE's subsurface contamination problems in Chapters 2 and 5, the committee recommends that the subsurface component of the EM Science Program have the following four research emphases:

The committee recommends that the subsurface component of the EM Science Program have the following four research emphases: 1. Location and characterization of subsurface contaminants and characterization of the subsurface. . . . 2. Conceptual modeling. . . . 3. Containment and stabilization. . . . 4. Monitoring and validation.

1. *Location and characterization of subsurface contaminants and characterization of the subsurface.* Basic research that supports advances in capabilities to locate and characterize subsurface contamination and elucidate relevant subsurface conditions will help DOE to better assess the potential hazards of its contamination problems and to design and implement appropriate corrective action strategies. Moreover, research on subsurface heterogeneity in geology, geochemistry, hydrology, and microbiology will provide a framework for assessing the fate and transport of contaminants. The committee believes that basic research is needed to support the development of the following capabilities:

   - improved capabilities for characterizing the physical, chemical, and biological properties of the subsurface;
   - improved capabilities for characterizing physical, chemical, and biological heterogeneity, especially at the scales that control contaminant fate and transport behavior;
   - improved capabilities for measuring contaminant migration and system properties that control contaminant movement;
   - methods to integrate data collected at different spatial and temporal scales to better estimate contaminant and subsurface properties and processes; and
   - methods to integrate such data into conceptual models.

2. *Conceptual modeling.*[3] Basic research on the fundamental approaches and assumptions underlying conceptual model development could produce a "tool box" of methodologies that can be applied to contaminated sites both inside and outside the DOE complex. This research should focus on the following topics:

[3]A conceptual model is a description of the subsurface as estimated from knowledge of the known site geology and hydrology and the physical, chemical, and biological processes that govern contaminant behavior. See Chapter 5.

- new observational and experimental approaches and tools for developing conceptual models that apply to complex subsurface environments;
- new approaches for incorporating geological, hydrological, chemical, and biological subsurface heterogeneity into conceptual model formulations at scales that dominate flow and transport behavior;
- development of coupled-process models through experimental studies at variable scales and complexities that account for the interacting physical, chemical, and biological processes that govern contaminant fate and transport behavior;
- methods to integrate process knowledge from small-scale tests and observations into model formulations;
- methods to measure and predict the scale dependency of parameter values; and
- approaches for establishing bounds on the accuracy of parameters and conceptual model estimates from field and experimental data.

3. *Containment and stabilization.* Increasing reliance is being placed on containment and stabilization because DOE recognizes that cleanup at some sites is technically infeasible, or that contamination at some sites does not pose a high risk to humans or the environment. Basic research that supports the development of new waste containment and stabilization technologies could lower the cost, accelerate regulatory approvals, and increase public confidence in solving subsurface contamination problems. Research focused on the following topics is especially needed:

    - mechanisms and kinetics of chemically and biologically mediated reactions that can be applied to new stabilization and containment approaches or that can be used to understand the long-term reversibility of chemical and biological stabilization methods;
    - physical, chemical, and biological reactions that occur among contaminants, soils, and barrier components so that more compatible and durable materials for containment and stabilization systems can be developed;
    - fluid transport behavior in conventional barrier systems; and
    - development of methods for assessing the long-term durability of containment and stabilization systems.

4. *Monitoring and validation.* Basic research leading to improvements in capabilities to monitor and validate contaminant locations and perform remedial actions will greatly enhance the technical success of DOE's efforts to remediate or contain and stabilize contamination. Many of the research opportunities for monitoring and validation have been covered in the research emphases discussed above. In addition, the committee believes that basic research is needed on the following topics:

   - development of methods for designing monitoring systems to detect both current conditions and changes in system behaviors;
   - development of validation processes.
   - determining the key measurements that are required to validate models and system behaviors, the spatial and temporal resolutions at which such measurements must be obtained, and the extent to which surrogate data can be used in validation efforts; and
   - research to support the development of methods to monitor fluid and gaseous fluxes through the unsaturated zone, and for differentiating diurnal and seasonal changes from longer-term secular changes.

Within these four emphases, the committee further recommends that the EM Science Program encourage research on metals and radionuclides, which is generally not receiving much attention in other federal research programs. There should, however, be sufficient flexibility in the program so that support can be provided for high-risk but potentially high-payoff research ideas that intersect with these recommended research emphases.

The committee's recommendation of these four research emphases does not mean that the subsurface research in the current program portfolio is inappropriate or misdirected. Rather, the recommended emphases represent areas where more research clearly is needed.

Implementation

The EM Science Program is a basic research program focused on very real DOE problems. The program's success will be measured both by its impact on advancing the science needed for site remediation and its impact on DOE site cleanup. To be successful, the program must not only be focused on the right problems but it also must encourage researchers to do the right work; and it must be structured so that research results can be handed off to technology developers and problem holders at DOE sites. The committee concluded that the following actions would help ensure the long-term success of the program in meeting the first two of these objectives:[4]

1. *Program Integration.* Program managers must encourage and support program-wide integration activities to optimize impacts of advances in subsurface science on DOE site cleanup. To this end, the program's implementation strategy should contain the following integrative elements:

   - Continue to reach beyond the usual group of DOE researchers to pull in new and novel ideas to address DOE-specific problems.
   - Continue to encourage multidisciplinary research and university-national laboratory-industry collaborations that will promote new insights into the very complex subsurface problems at DOE sites.
   - Integrate existing data and ideas—both from DOE sites and basic research programs outside DOE—to promote advancements in subsurface science and improvements in capabilities for addressing DOE's subsurface contamination problems.

2. *Field Sites.* The committee recommends that program managers examine the feasibility of developing field research sites as one program component. Such sites could attract new researchers to the program, encourage both formal and informal multidisciplinary collaborations among the researchers, and facilitate the transfer of research results into application. These field sites could include contaminated or uncontaminated areas at major DOE sites; analog uncontaminated sites that have subsurface characteristics similar to those at contaminated DOE sites; and even virtual sites comprised of data on historical and contemporary contamination problems. These sites could be established by the program itself or in cooperation with other research programs.

   The establishment of field research sites is potentially expensive, especially if the sites are located in contaminated areas. Consequently, the establishment of such sites will require additional budget support beyond that required to fund individual

[4]The third objective on moving science into application, although extremely important, is beyond the statement of task for the present study.

research projects, and well beyond the amount of funding available to the program for new starts in fiscal year 1999. Moreover, the use of such sites will have to be evaluated periodically to determine whether they are adding value to the research effort, particularly given the cost of such sites relative to the total size of the program budget.

3. *Program Funding.* The issue of program funding has received a great deal of attention from a previous NRC committee (NRC, 1997b), which concluded that the "program must be large enough to support a significant number of 'new starts' (i.e., new projects or competitive renewals) each year if it is to be successful in attracting innovative proposals from outstanding researchers ...." New starts will help establish a cadre of knowledgeable and committed investigators—undergraduates, graduates, postdocs, and professionals—who can be called on by DOE in the years ahead for help with its most difficult contamination problems. New starts also are needed to maintain continuity in the research effort since the advancement of scientific knowledge is a cumulative effort involving many scientists over long periods of time. This effort is set back significantly each time program funding is interrupted.

   It is the committee's strong impression that the current level of program funding is not sufficient to support the research emphases outlined in this report, especially since subsurface research is just one of many research areas supported by the program. The committee has no basis on which to recommend a specific funding level, and such a recommendation would be well beyond the committee's statement of task. The committee believes that it is the responsibility of program managers to estimate the amount of funding required to provide adequate support for a research program focused on the knowledge gaps presented in Chapter 5. One approach for estimating the annual budget needed to support the recommended research is to estimate the number of projects needed to attain a critical mass of research on each technical challenge area discussed earlier, and then to multiply that number by the average annual grant size. Such estimates could be used to justify future and possibly larger budget requests to upper DOE management and Congress, especially if the estimates are reviewed and validated by DOE's internal and external advisory committees. Future budget requests are likely to be seen in an increasingly more favorable light as the program becomes more firmly connected to EM's cleanup problems.

It is the committee's strong impression that the current level of program funding is not sufficient to support the research emphases outlined in this report ....

Concluding Remarks

There must be strong scientific, technical, and management leadership at all levels ... if significant progress on closing the knowledge gaps is to be made in the next decade and the research results are to be applied effectively to the DOE cleanup program.

The basic research supported by the EM Science Program and other relevant federal research programs will have little if any impact on DOE cleanup unless research results are transferred into technology development programs in EM and to problem holders at DOE sites. Program managers have a responsibility to ensure that the handoff from research to development is timely and effective, both for research results developed in its programs and from other relevant federal programs.

There must be strong scientific, technical, and management leadership at all levels, from the EM Science Program up to and including the assistant secretary for environmental management if significant progress on closing knowledge gaps is to be made in the next decade and the research results are to be applied effectively to the DOE cleanup program. The development of this leadership is a continuing challenge—and a significant opportunity—for the EM Science Program and DOE.

1

Introduction and Task

The Department of Energy's (DOE's) Environmental Management (EM) Science Program was created by the 104th Congress[1] to bring the nation's basic science infrastructure to bear on the massive environmental cleanup effort now underway in the DOE complex. The objectives of the program are to

- provide scientific knowledge that will revolutionize technologies and cleanup approaches to significantly reduce future costs, schedules, and risks;
- bridge the gap between broad fundamental research and needs-driven applied technology; and
- focus the nation's science infrastructure on critical DOE environmental management problems.

To meet these objectives, the EM Science Program provides three-year awards to investigators in industry, national laboratories, and universities to undertake research on problems relevant to DOE cleanup efforts. Project awards are competitive and are made on the basis of merit and relevance reviews managed through a partnership between the DOE Office of Environmental Management, which has the primary responsibility for the cleanup mission, and the DOE Office of Science,[2] which manages DOE basic research programs. A more detailed description of the program is provided in Appendix A.

Since its establishment by Congress, the program has held four proposal competitions and has awarded about $225 million in funding, which puts it among the largest environmental research efforts in the federal government (see Chapter 4). The first two proposal competitions

[1]Public Law 104-46, 1995.

[2]Formerly the Office of Energy Research.

were completed in fiscal years 1996 and 1997 and resulted in 202 awards totaling about $160 million. These awards covered a wide range of problems related to cleanup of the defense complex, including subsurface contamination problems.[3] The third proposal competition was completed in fiscal year 1998 and resulted in 30 awards totaling about $30 million. These awards provided funding for projects primarily related to high-level radioactive waste and deactivation and decommissioning. The fourth proposal competition was completed in fiscal year 1999, while this report was in the end stages of completion, and focused primarily on subsurface contamination and low dose radiation.[4]

Shortly after the program was established, DOE requested advice from the National Academies on its structure and management. In response, the National Academies established the Committee on Building an Effective Environmental Management Science Program, which operated from May 1996 through March 1997 and produced three reports.[5] One of the primary recommendations made by this committee was that DOE should

> *develop a science plan for the EMSP [Environmental Management Science Program]. This science plan should provide a comprehensive list of significant cleanup problems in the nation's nuclear weapons complex that can be addressed through basic research and a strategy for addressing them. (NRC, 1997b, p. 3)*

This committee also recommended a near-term and a long-term process for developing this science plan: For the near term, program managers should develop a science plan from existing DOE documents. For the longer term, DOE should consult with its problem holders (i.e., site technical staff, managers, and stakeholder advisory groups who have knowledge of the cleanup issues) about cleanup problems that cannot be resolved practically or efficiently with current knowledge or technologies.

---

[3]An analysis of the program's subsurface science portfolio for fiscal years 1996 and 1997 is provided in Chapter 3.

[4]Thirty-one awards totaling $25 million were made for projects related to subsurface contamination research, and eight awards totaling about $8 million were made for low dose radiation research in cooperation with the DOE Office of Science's Low Dose Radiation Research Program. The committee did not have an opportunity to review the fiscal year 1999 projects.

[5]*Building an Effective Environmental Management Science Program: Initial Assessment* (NRC, 1996a); *Letter Report on the Environmental Management Science Program* (NRC, 1996b); and *Building an Effective Environmental Management Science Program: Final Assessment* (NRC, 1997b). All three reports can be viewed at the National Academy Press Web site (http://books.nap.edu/catalog/5557.html).

SIDEBAR 1.1 STATEMENT OF TASK

The objective of this study is to provide recommendations to DOE's EM Science Program on the formulation of a long-term basic research[1] program to address subsurface contamination problems at DOE sites. These recommendations will take into account significant subsurface contamination problems at major DOE sites that cannot be addressed with current technologies and science knowledge gaps relevant to these problems. The recommendations also will take into account the research already completed and currently in progress by other federal and state agencies and will identify areas of research where the EM Science Program can make significant contributions to address DOE's subsurface contamination problems and to add scientific knowledge generally.

[1]Scientific research comprises a spectrum of investigative activities that are frequently classified using artificial groupings such as basic and applied (e.g., Pielke and Byerly, 1998). In the committee's view, basic research is defined as research that creates new generic knowledge and is focused on long-term, rather than short-term, problems. See also NRC (1995).

In the spring of 1998, Gerald Boyd, the then-acting director (now director) of the Office of Science and Technology, requested that the National Academies convene another committee of experts to advise DOE on its first science plan for the EM Science Program, which DOE had decided would address subsurface contamination. In response, the current committee was formed under the joint auspices of the Board on Radioactive Waste Management and Water Science and Technology Board. This committee has expertise in basic research and research management in the scientific disciplines relevant to subsurface contamination problems at DOE sites.[6]

The statement of task for this study (see Sidebar 1.1) charged the committee to provide recommendations for a science research program for subsurface contamination problems at DOE sites, and especially to identify areas of research where the program could make significant contributions to DOE's cleanup efforts and add to scientific knowledge generally. The committee held six meetings between October 1998 and July 1999 to gather information on subsurface contamination and related problems at six major DOE sites and to develop this report.[7] The committee also produced an interim report to advise DOE on the fiscal year 1999 proposal call. That report is given in Appendix E.

The committee received briefings on subsurface contamination problems at the Hanford Site (Washington), Idaho National Engineering and Environmental Laboratory, Nevada Test Site, Oak Ridge Site

[6]Biographical sketches of committee members are given in Appendix C.
[7]See Appendix B for a summary of the information-gathering activities.

(Tennessee), and Savannah River Site (South Carolina). The committee toured the Hanford Site and Savannah River Site to make direct observations of the problems and obtain briefings from site personnel, and it reviewed DOE and other documents concerning the subsurface contamination problems at these sites and at the Rocky Flats Site in Colorado. The committee did not request briefings on the Rocky Flats Site because of time constraints and because DOE advised that its planned cleanup activities of this site would be completed by 2006 (e.g., DOE, 1998a).

The committee focused primarily on the scientific issues in keeping with its collective basic-research expertise. The committee has reviewed the subsurface contamination problems at major DOE sites (see Chapter 2) and provides recommendations on a research agenda to address these problems (see Chapter 5). The committee also considered the research being sponsored by other federal programs (see Chapter 4) as well as the projects supported in the current EM Science Program portfolio (see Chapter 3), so that unnecessary duplication of effort can be minimized.

In Chapter 6, the committee recommends a strategy for implementing a research agenda, but it has refrained from making recommendations on program management, which is largely beyond its collective expertise and was covered in detail by a previous National Academies committee (NRC, 1997b). The committee also comments on the level of effort (both in time and funding) that will be required to make significant progress on the research agenda. The committee believes that the success of the EM Science Program will depend both on the nature of the problems addressed and on the effort sustained in solving them.

2

Subsurface Contamination in the DOE Complex

Over the last five decades, the United States has created a massive industrial complex to develop, test, manufacture, and maintain nuclear weapons for national security purposes. The U.S. Army Corps of Engineers, Manhattan Engineering District, started constructing the complex during the Second World War. The complex was expanded during the ensuing Cold War by the Atomic Energy Commission, the Energy Research and Development Authority, and starting in 1977, the Department of Energy (DOE). The DOE complex, as it has come to be known, encompasses 134 distinct geographic sites in 31 states and one territory with a total area of over two million acres (DOE, 1998a). The individual sites range in size from several hundred square miles to less than one square mile; these sites host a variety of defense-related activities ranging from uranium mining and milling to nuclear weapons testing (see Figure 2.1).

The production and testing of nuclear weapons has created a legacy of significant environmental contamination, as described in some detail later in this chapter. In 1989, Congress created the Office of Environmental Management (EM) in DOE to reduce threats to health and safety posed by the environmental contamination at DOE sites. To meet this objective, EM has undertaken a major cleanup effort, which, according to DOE, is the largest environmental cleanup in the world. This is certainly true from a cost standpoint: EM is now spending about $5.8 billion per year on its cleanup program and has spent over $50 billion since 1990. It expects to spend another $147 billion between 1997 and 2070 (DOE, 1998a), but this estimate is uncertain because the magnitude of contamination and the level of cleanup effort required at some sites are still poorly understood.

In this chapter, the committee provides an overview of the subsurface contamination problems around the DOE complex and shows by

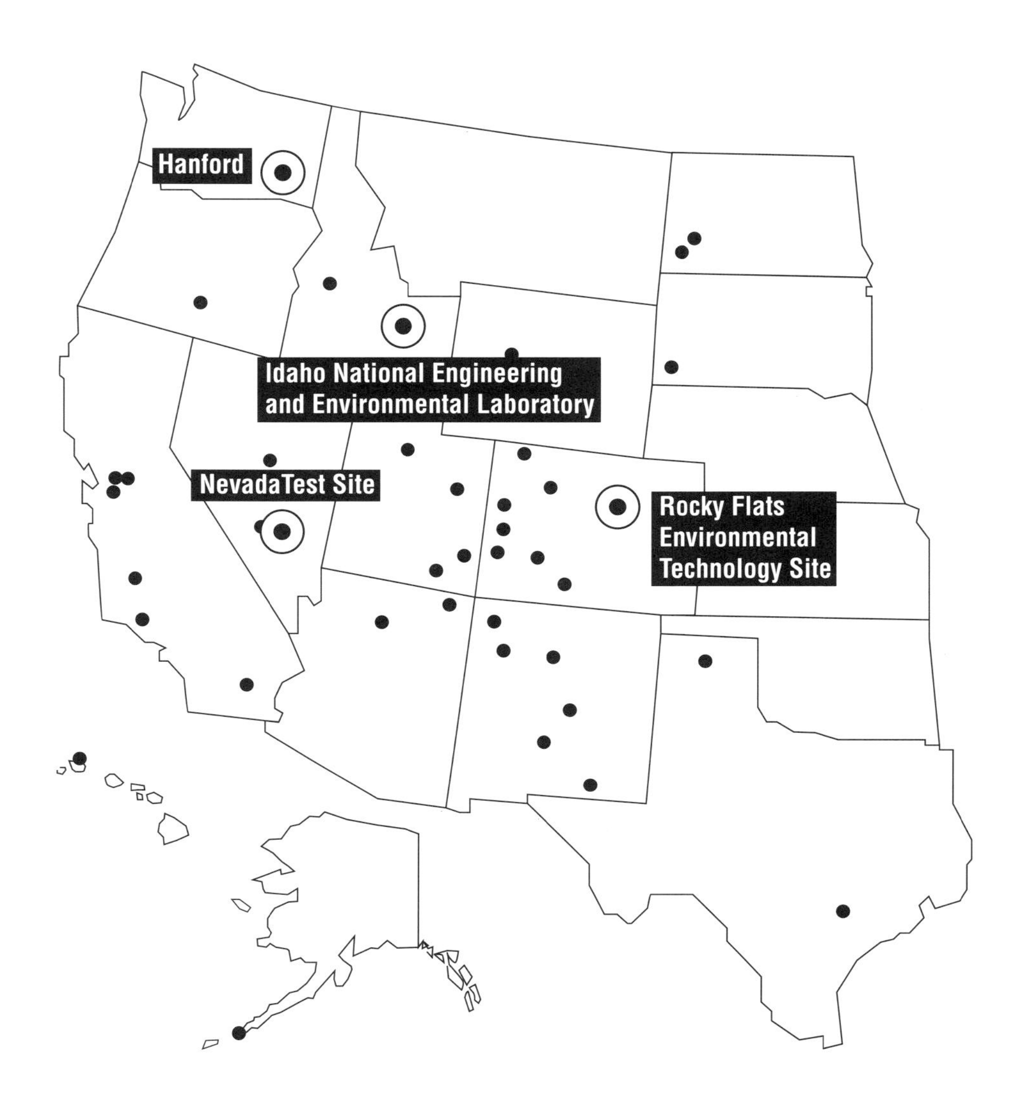

FIGURE 2.1 Location of DOE complex sites. The major sites are labeled by name on the figure. The locations of other sites are indicated by closed circles. SOURCE: DOE.

example how advances in scientific and engineering knowledge can improve cleanup effectiveness. The chapter is organized into three sections. The first provides an overview of the DOE complex and its mission and describes the legacy of contamination from weapons production and related activities. The second section illustrates the range of subsurface problems that exist across the complex today and what DOE is doing to correct them. The examples are taken from the six largest DOE sites: Hanford, Idaho, Nevada, Oak Ridge, Rocky Flats, and Savannah River (see Sidebar 2.1). In the third section, the committee discusses how scientific and engineering research can improve the effectiveness of DOE's mission to tackle these contamination problems.

This discussion will be used to support the recommendations in Chapters 5 and 6.

Past Practices and Consequences

Nuclear weapons production during the Cold War was a highly industrialized enterprise that involved a vast complex of mines and industrial sites across the United States. The front end of the process was focused on the production of uranium, which was then used to produce other weapons materials, particularly plutonium and tritium.

SIDEBAR 2.1 THE DOE COMPLEX

Although the DOE complex encompasses over 100 distinct sites, much of the major defense-related activities were conducted at the six largest DOE sites (see Figure 2.1) described below.

The *Hanford Site* is located in southeastern Washington state and covers an area of about 1,450 square kilometers (560 square miles). Production of materials for nuclear weapons took place here from the 1940s until mid-1989. The site contains several production reactors, chemical separations plants, and solid and liquid waste storage sites.

The *Idaho National Engineering and Environmental Laboratory,* first established as the Nuclear Reactor Testing Station and then the Idaho National Engineering Laboratory, occupies 2,300 square kilometers (890 square miles) in a remote desert area along the western edge of the upper Snake River plain. The site was established as a building, testing, and operating station for various types of nuclear reactors and propulsion systems, and the site also manages spent fuel from the naval reactor program.

The *Nevada Test Site,* which occupies about 3,500 square kilometers (1,350 square miles) in southern Nevada, was the primary location for atmospheric and underground testing of the nation's nuclear weapons starting in 1951.

The *Oak Ridge Reservation* covers an area of approximately 155 square kilometers (60 square miles) and is located about 10 kilometers (6 miles) west of Knoxville, Tennessee. The reservation has three major operating facilities: the Oak Ridge National Laboratory, the Y-12 Plant, and the K-25 Plant. The laboratory was originally constructed as a research and development facility to support plutonium production technology. The Y-12 Plant was built to produce highly enriched uranium by electromagnetic separation; and the K-25 Plant, formerly known as the Oak Ridge Gaseous Diffusion Plant, also was created to produce highly enriched uranium for nuclear weapons.

The *Rocky Flats Environmental Technology Site* is situated on about 140 hectares (~350 acres) near Denver, Colorado, and has more than 400 manufacturing, chemical processing, laboratory, and support facilities that were used to produce nuclear weapons components. Production activities once included metalworking, fabrication and component assembly, and plutonium recovery and purification. Operations at the site ceased in 1989.

The *Savannah River Site,* located near Aiken, South Carolina, covers an area of about 800 square kilometers (300 square miles). The site was established in 1950 to produce special radioactive isotopes (e.g., plutonium and tritium) for use in the production of nuclear weapons. The site contains production reactors, chemical processing plants, and solid and liquid waste storage sites.

The back end was focused on the fabrication and testing of nuclear devices. The major production steps and waste byproducts are described in Sidebar 2.2.

The United States is no longer producing plutonium and tritium[1] for

---

[1]The secretary of energy has announced that DOE may produce tritium in the future to replenish current stocks of nuclear weapons.

TABLE 2.1 Principal Dense Non-Aqueous Phase Liquid (DNAPL), Metal, and Radionuclide Contaminants in the DOE Complex

DNAPLs	Metals	Radionuclides
Trichloroethylene	Lead	Plutonium
Dichloroethylene	Chromium (VI)	Strontium-90
Tetrachloroethylene	Mercury	Cesium-137
Perchloroethylene	Zinc	Uranium (various isotopes)
Chloroform	Beryllium	Tritium
Dichloromethane	Arsenic	Thorium
Polychlorinated Biphenyls	Cadmium	Technetium-99
	Copper	Radium
		Iodine-129

SOURCE: EPA (1977); INEEL (1997); Riley and Zachara (1992).

nuclear weapons, and a large part of the DOE complex has been shut down or placed on standby. All of DOE's production reactors have been shut down, and only two reprocessing facilities (the F and H canyons at Savannah River) continue to operate. These are scheduled to be phased out during the next decade. The weapons design and assembly facilities also continue to operate, but their mission now includes the disassembly of surplus nuclear weapons. The Nevada Test Site remains open, but only subcritical nuclear tests have been conducted there since 1992.

During the last decade, a large part of the DOE complex, including some of the sites discussed in Sidebar 2.1, have taken on a new mission: namely, remediation of the environmental contamination resulting from weapons production. This mission is formidable, because it involves cleanup of a wide variety of hazardous chemicals and radioactive materials introduced into the environment during five decades of weapons production and testing (see Sidebar 2.3). The contaminants include dense non-aqueous phase liquids (DNAPLs; see Sidebar 2.4); toxic metals such as lead, chromium, and mercury; and radionuclides such as plutonium, cesium, strontium, and tritium (see Table 2.1).

These contaminants were introduced into the environment through a variety of pathways, including intentional disposal into the ground through injection wells, disposal pits, and settling ponds; and through accidental spills and leaks from storage tanks and waste transfer lines. In some cases, there is little information available on either the timing or magnitude of contaminant releases to the environment, or the fate of contaminants in the subsurface after release. Moreover, DOE sites are

SIDEBAR 2.2 NUCLEAR FUEL CYCLE AND NUCLEAR WEAPONS PRODUCTION

The production of nuclear weapons is a technically complex and highly industrialized process. The major production steps and waste byproducts of this process are described below.

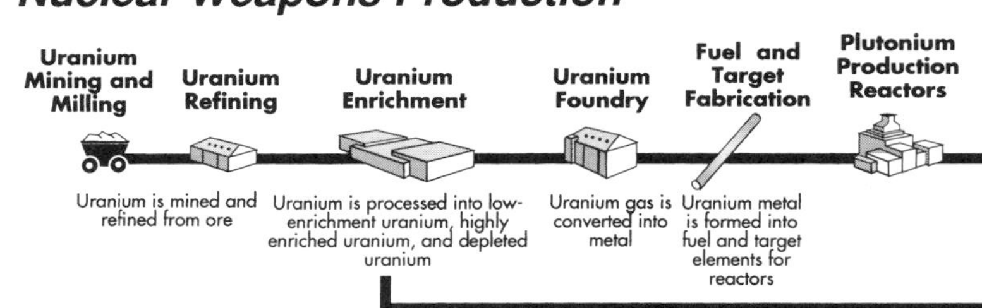

Mining and milling. **Uranium ore was mined at over 400 sites in the United States and processed in mills to produce uranium oxide. These processes produced large volumes of mine and mill tailings that contained heavy metals and radioactive radium and thorium. This waste is being managed through the Uranium Mill Tailings Radiation Control Act program.**

Uranium enrichment. **Elaborate chemical processes were used to concentrate the fissile isotope uranium-235 from the milled ore. Uranium enrichment facilities were built at Oak Ridge (Y-12 and K-25 Plants), Ohio (Portsmouth Plant), and Kentucky (Paducah Plant). The waste streams from the enrichment process include depleted uranium (i.e., depleted in U-235 relative to U-238), uranium-contaminated scrap metal, polychlorinated biphenyl-contaminated waste, and a variety of organic solvents. Separation of lithium isotopes at the Oak Ridge Y-12 plant also produced large amounts of mercury waste.**

Fuel and target fabrication. **The enriched uranium was converted to metal at the Fernald Plant in Ohio and then fabricated into reactor fuel or targets for plutonium production at Hanford and Savannah River. These processes produced uranium dust and a variety of chemical wastes.**

Plutonium production. **The United States produced about 100 metric tons of plutonium between 1944 and 1988 at 14 reactors at the Hanford and the Savannah River sites. The reactors at Savannah River also produced tritium. Thousands of tons of uranium fuel were processed through the reactors during their four decades of operation. The waste streams from these operations include solid and liquid**

located in a variety of climatic zones and have complex subsurface characteristics (see Table 2.2), which makes it difficult to predict the location, transport, and fate of contaminants once they are released into the environment. As discussed in some detail in other National Research Council reports (NRC, 1997a, 1999), technologies to effectively remediate many subsurface DNAPL, metal, and radionuclide contamination problems are either lacking or are unproven for large-scale site remediation.

Although subsurface contamination is generally acknowledged to be a significant problem across the DOE complex, estimates of the magni-

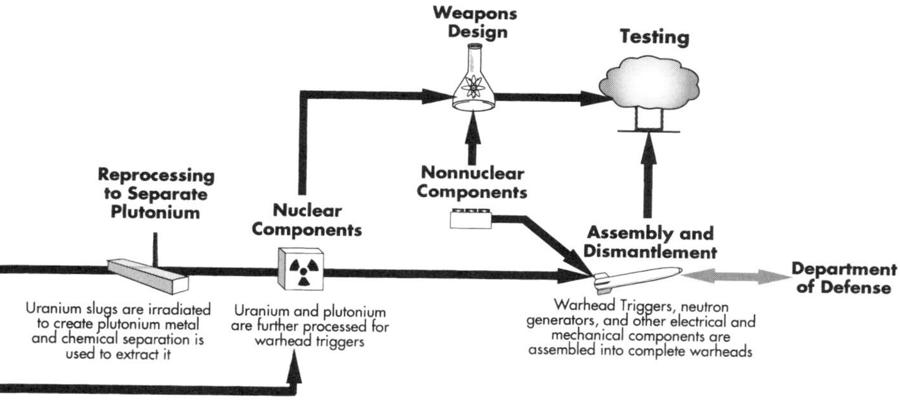

Figure Source: DOE

radioactive waste, acids, and solvents. The cooling water from the reactors contained some radionuclides, most notably tritium.

Plutonium Separation. **Plutonium and other special isotopes were separated from the irradiated fuel by a variety of chemical processes. Chemical separations plants were located at the Hanford, Savannah River, and Idaho sites. Operation of the separations plants produced significant volumes of highly radioactive and hazardous chemical waste and water containing low levels of radionuclides and hazardous chemicals.**

Weapons design, fabrication, and assembly. **Weapons design was the responsibility of the Los Alamos and Lawrence Livermore National Laboratories. Weapons components were produced at several sites in the United States, and final assembly took place at the Pantex Plant in Texas. The fabrication process produced several waste streams, including scrap uranium and plutonium metal and solvents.**

Weapons testing. **The United States has conducted more than a thousand nuclear weapons tests in the atmosphere, under water, and underground, and most have occurred at the Nevada Test Site. This testing resulted in the contamination of surface and subsurface sites with radioactive materials, including tritium, plutonium, and fission products.**

tude of the problem vary considerably, as shown in Table 2.3. According to recent DOE estimates (DOE, 1998a) there are about 6.4 billion cubic meters (226 billion cubic feet) of contaminated soil, groundwater, and related environmental media at its sites.[2] Most of this contamina-

[2]The subsurface contamination estimates provided in this chapter are compiled from various DOE documents. The committee cannot evaluate the accuracy of any of these estimates, but believes based on the briefings and documents it received during the course of this study that the estimates are likely to have very large uncertainties.

tion is at two sites, the Hanford Site in eastern Washington and the Idaho National Engineering and Environmental Laboratory in south-central Idaho (see Figure 2.1). At these two sites alone, EM cleanup is not expected to be completed before 2050, and after cleanup is "complete" EM does not know how much contamination will remain in the ground to be managed through surveillance and containment.

EM's current cleanup plans, which also are given in the *Paths to Closure* report (DOE, 1998a), anticipate expenditures on the order of $57 billion between 1997 and 2006 to complete cleanup at all but 10

TABLE 2.2 Geologic and Climatologic Variability Across the DOE Weapons Complex

DOE Site	Climate	Geology and Hydrogeology	Surface Waters	Depth to Groundwater (m)
Savannah River Site	Humid, subtropical	Atlantic Coastal Plain with clay soils. The strata are deeply dissected by creeks, and most groundwater eventually seeps into and is diluted by the creeks.	Savannah River	0-38[a]
Hanford Site	Arid, cool; mild winters and warm summers; average annual rainfall 16 cm (6.3 in.)	Alluvial plain of bedded sediments with sands and gravels. Groundwater flows toward the Columbia River.	Columbia River	60-90[b]
Oak Ridge Reservation	Humid, typical of the southern Appalachian region; average annual precipitation 138 cm (54.4 in.)	Valley and ridge province bordering the Cumberland Plateau. Primary porosity is low, but fracture porosity is present. High clay content. Shallow water table.	Clinch River	6-37[c]
Rocky Flats Environmental Technology Site	Temperate, semiarid, and continental temperatures; average annual rainfall just under 40 cm (15 in.)	Colorado Piedmont section of the Plains physiographic province. Alluvial deposits cover the site.	Several streams occur on or near the facility	0-9[d]
Idaho National Engineering and Environmental Laboratory	Semiarid with sagebrush-steppe characteristics located in a belt of prevailing western winds; average annual rainfall 22 cm (8.5 in.)	Near the northern margin of the Eastern Snake River plain, a low-lying area of late Tertiary and Quaternary volcanism and sedimentation. Basalt covers three-quarters of its surface.	Big Lost River and other ephemeral streams	60-240

[a]Michelle Ewart, SRS, personal communication, 2000.
[b]Gephart and Lundgren (1998).
[c]Grover Chamberlain, DOE-HQ, personal communication, 2000.
[d]Christine Gelles, DOE-HQ, personal communication, 2000.

SOURCE: Adapted from Sandia National Laboratories (1996), except where noted.

of its sites, including the major sites shown in Table 2.3. DOE expects an additional expenditure of $79 billion to clean up those remaining 10 sites between 2007 and 2070. About $14 billion will be incurred for

SIDEBAR 2.3 A PRIMER ON RADIOACTIVE WASTE

Radioactive wastes are the unwanted byproducts of the nuclear fuel cycle (see Sidebar 2.2) and may contain both radioactive isotopes and hazardous chemicals. In the United States, radioactive waste is classified and managed by its source of production rather than by its physical, chemical, or radioactive properties. Consequently, different classes of waste can contain many of the same radioactive isotopes, and even "low-level" waste can contain certain long-lived radioactive isotopes.

In general, nuclear fuel cycle wastes are grouped into the following broad classes for purposes of management and disposal:

- ***Mill tailings*** **are wastes resulting from the processing of ore to extract uranium and thorium.**
- ***Spent nuclear fuel*** **is fuel that has been irradiated in a nuclear reactor, and for the purposes of disposal may include cladding and other structural components.**
- ***High-level waste*** **is the primary waste produced from chemical processing of spent nuclear fuel. This waste is usually liquid in form and contains a wide range of radioactive and chemical constituents. Spent nuclear fuel is often referred to as high-level waste in nuclear waste management terminology although it is defined differently in the regulations.**
- ***Transuranic waste*** **excludes high-level waste as defined above and includes waste that contains alpha-emitting transuranium (i.e., atomic number greater than 92) isotopes with half lives greater than 20 years and concentrations greater than 100 nanocuries per gram. DOE also includes U-233 in its definition of transuranic waste. This waste usually consists of contaminated materials like clothing and tools resulting from the manufacture of nuclear weapons.**
- ***Low-level waste*** **is radioactive waste that does not meet one of the definitions given previously.**

There are two other classes of materials that DOE sometimes manages as waste:

- ***Nuclear materials,*** **such as plutonium and special-use isotopes, that may be declared as surplus and disposed of as waste.**
- ***Contaminated environmental media,*** **such as contaminated soil and groundwater, that may fall under the Environmental Protection Agency's Comprehensive Environmental Response, Compensation and Liability Act. The cleanup of this contamination may generate additional radioactive and chemical waste streams that must be treated and managed.**

In the United States, the federal government regulates the management and disposal of most types of radioactive waste. Federal regulations seek to reduce to reasonably achievable levels the exposure of workers and other members of the public to this waste. The guiding philosophy for waste management is sequestration, that is, to isolate the waste from human populations and the environment, either through long-term storage or disposal in an underground facility until it no longer poses a hazard.

remedial action, which is defined by DOE as the characterization and cleanup of sites where contaminants or contaminated materials were released into the environment. The cleanup of these sites will involve the recovery and treatment of abandoned materials; remediation of soil, groundwater and surface water; and monitoring where contamination cannot be cleaned up to unrestricted release standards.

According to EM, site cleanup will be considered "complete" when, among other things, releases to the environment have been cleaned up in accordance with agreed standards and groundwater contamination

TABLE 2.3 Projected Magnitude, Timing, and Cost of DOE Cleanup Activities

DOE Site	Projected End State(s)[a]	Completion Date of Planned Cleanup Projects	Soil, Groundwater, and Other Media Requiring Remedial Action (10^6 m^3)	Pre-2006 Life-Cycle Costs (1998 $B)	Post-2006 Life-Cycle Costs (1998 $B)[b]	Residual Contaminants in Soil			Residual Contaminants in Water		
						Metals	Rads	Organics	Metals	Rads	Organics
Hanford	IM, other TBD	2046	1,400	13	37.4	●	●	●		●	●
Idaho	UR, RR, IM	2050	4,700	5.1	11.3	●	●	●	●	●	●
Nevada Test Site & Other Associated Sites[c]	RR, IM, other TBD	2014	3.1[d]	0.92	1.3		●	●		●	
Oak Ridge & Associated Sites[e]	UR, RR, IM	2013	31	5.4	7.7	●	●	●	●	●	●
Rocky Flats	UR, RR, IM	2006-2010	0.79	5.3	0.96	●	●	●		●	●
Savannah River	IM, other TBD	2038	172	12	17.7	●	●	●	●	●	●
Other Sites	UR, RR, IM, other TBD	1999-2038	120	7.8	2.8	●	●	●	●	●	●
Totals			6,400	50	79						

[a]UR = unrestricted release; RR = restricted release; IM = long-term institutional management; TBD = to be determined.

[b]Post-2006 cost estimates include some but not all costs for long-term institutional management.

[c]Includes the Nevada Test Site and eight off-site locations in five states (Alaska, Colorado, Mississippi, Nevada, and New Mexico) where underground nuclear tests were conducted.

[d]Estimate does not include groundwater contaminated by nuclear testing.

[e]Includes the Oak Ridge Reservation, the Paducah and Portsmith Gaseous Diffusion Plants in Kentucky and Ohio, respectively, and the Weldon Spring Site in Missouri.

SOURCE: Compiled from DOE (1998a, 1999).

has been contained or long-term treatment or monitoring has been put in place (DOE, 1998a, p. 1-7). **In other words, even after EM has completed its cleanup projects there will still be contaminants left in the subsurface and in surface land-disposal facilities that will require long-term management and possibly future actions to prevent further spread.**

Examples of Subsurface Contamination Problems at Major DOE Sites

The committee received several briefings on soil and groundwater contamination problems and remediation activities at five of the six major DOE sites (see Sidebar 2.1): Idaho, Hanford, Nevada, Oak Ridge,

SIDEBAR 2.4 NON-AQUEOUS PHASE LIQUIDS IN HETEROGENEOUS FORMATIONS

Non-aqueous phase liquids (NAPLs) are a common class of subsurface contaminants at many DOE sites. Dense non-aqueous phase liquids (or DNAPLs) are organic chemicals such as trichloroethylene, tetrachloroethylene, and polychlorinated biphenyls that have densities greater than water (i.e., > 1.0 gram per cubic centimeter) at standard temperature and pressure and have low solubilities. Their relatively high density causes them to migrate downward through soils and groundwater under the influence of gravity. When they encounter a low-permeability layer, they may pool or move laterally. Because of their low solubilities, NAPLs remain as a separate phase and may provide a long-term source of groundwater contamination.

The detection, characterization, and remediation of DNAPL contamination is generally difficult for a number of reasons, including geological heterogeneity; complex physical, chemical, and biological interactions; lack of efficient and cost effective field characterization techniques; and limitations and unavailability of properly validated modeling tools for the design and evaluation of remediation techniques. Experimental studies (e.g., Schwille, 1988; Kueper and Frind, 1991; Illangasekare and others, 1995) have shown that geologic heterogeneity can cause lateral spreading, preferential flow, and DNAPL pooling. In fact, such heterogeneities may be the major factor in controlling the entrapment distribution of DNAPLs in the subsurface. The DNAPL may exist as discontinuous, stable pore-scale masses trapped in soils under capillary forces, but it may also exist as an immobile continuous phase trapped by various heterogeneity features.

Researchers (e.g., Pfannkuch, 1984; Schwille, 1988) have identified two geometries associated with subsurface DNAPL contamination: (1) cylinders or fingers, and (2) pools on impermeable layers or bedrock. The experimental work by Illangasekare and others (1995) and the conceptual studies by Hunt and others (1986a,b) demonstrate that other geometries are possible as well, including zones of high saturation trapped in coarse lenses below the water table; thin pools trapped in coarse sand layers; and suspended pools trapped on top of fine sand or clay layers.

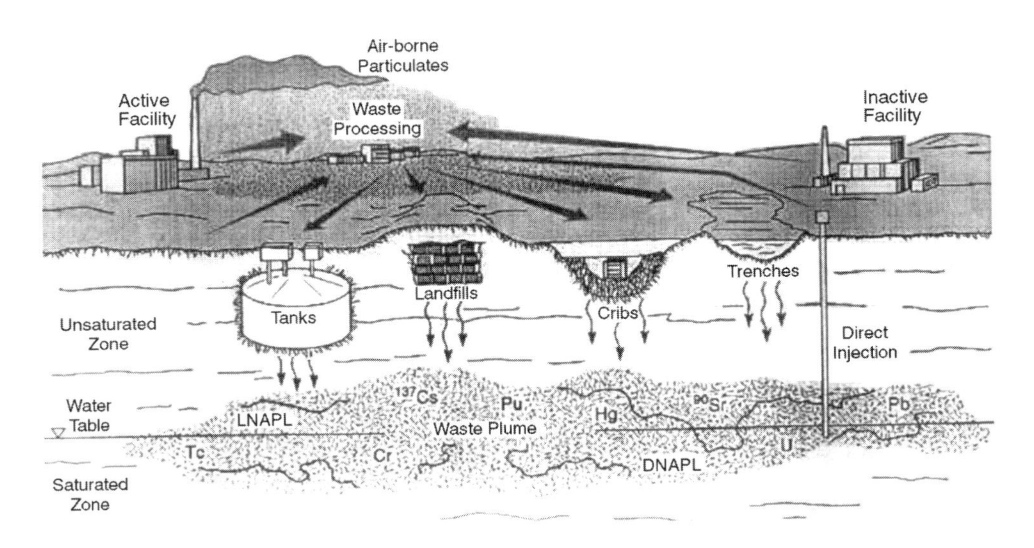

FIGURE 2.2 Schematic illustration of historical waste management practices in the DOE complex and contaminant pathways to the environment. SOURCE: DOE.

and Savannah River.[3] These sites are in different parts of the country (see Figure 2.1), are characterized by a wide range of geological and climatic conditions (see Table 2.2), and have a wide range of contamination histories.

In this section, the committee presents a snapshot of some of the sites' subsurface contamination problems to illustrate both the range of contamination problems and the remediation challenges. These examples are illustrative and do not necessarily represent the only significant contamination problems at the sites or across the DOE complex. Readers who wish additional information should consult the references cited in this section as well as the references given in Appendix D.

As will be shown in the following discussion, there are many similarities among the contamination problems at the major DOE sites. To highlight this fact, the committee has organized the discussion around different contaminant settings: waste burial ground contamination, soil contamination, unsaturated zone contamination, and saturated zone contamination. These are illustrated schematically in Figure 2.2.

Waste Burial Grounds

"Waste burial ground" is applied rather loosely to a wide array of disposal sites around the complex, ranging from auger holes to disposal pits and trenches. Waste burial grounds were used at all the major DOE

---

[3]As noted in Chapter 1, the committee did not obtain a briefing on the Rocky Flats site because of time constraints and because of DOE's plans to complete site cleanup by 2006. However, one of the committee members was familiar with the site, and the committee was able to obtain additional written information to develop the example used in this chapter.

sites to dispose of solid and liquid wastes, with many disposal practices now considered unacceptable by today's standards (see Sidebar 2.5): pits and trenches were unlined and frequently unmarked after closure and little thought was given to the stability or durability of waste that went into them. Consequently, there has been significant leakage from many waste burial grounds in the DOE complex, contaminating groundwater and surface water with metals, radionuclides, and hazardous chemicals. Efforts are now being made at some sites to excavate and remove the contaminants from these burial grounds or to cover them with low-permeability barriers to inhibit the further spread of contamination.

Burial Ground Complex at Savannah River

The Burial Ground Complex covers an area of about 80 hectares (195 acres) in the central part of the Savannah River Site and was used between 1952 and 1995 to dispose of low-level radioactive waste, mixed waste (i.e., radioactive and chemical waste), and intermediate-level radioactive wastes (see Plate 1). Contamination from these burial grounds has leaked to the underlying groundwater, producing four plumes consisting of various chemicals, metals, and radionuclides. The Burial Ground Complex represents one of the Savannah River Site's highest long-term risks to human health and environment and has been identified by the site's restoration division as its highest cleanup priority (Westinghouse Savannah River Co., 1998).

Plans to remediate this site have not been finalized but they will probably include several actions, including the removal or stabilization of highly contaminated zones in the southern part of the burial ground; installation of a multilayer surface barrier or cap consisting of natural and synthetic materials to impede water infiltration (see Plate 1); and long-term surveillance. DOE has relatively little experience with long-term caps, covers, and monitoring, but these containment approaches, if successful, are likely to find wide application for stabilization of waste burial grounds around the complex.

Radioactive Waste Management Complex at Idaho

The Radioactive Waste Management Complex was established in 1952 for disposal of solid low-level radioactive waste generated on site. Waste from other DOE sites was also buried here, including transuranic waste from Rocky Flats. After 1970, shallow land disposal of transuranic waste was discontinued, and above-ground storage on asphalt pads began to be used. Wastes were disposed in pits, trenches, soil vaults, an above-ground disposal pad, a transuranic storage area release site, and three septic tanks (DOE, 1996).

The Idaho site is located in a semiarid environment and is underlain by a thick unsaturated zone (see Table 2.2), which was thought to provide a barrier to contaminant migration to the underlying groundwater. However, low levels of plutonium have been found in groundwater beneath the Radioactive Waste Management Complex, and recent modeling work suggests that contaminant travel times to groundwater are only on the order of a few decades (see Sidebar 2.6), much shorter than anticipated when the complex was established in the 1950s.

One of the trenches contained in the complex is Pit 9, a one-acre site that was used for waste disposal primarily from Rocky Flats between 1967 and 1969. DOE estimates that Pit 9 contains about 7,100 cubic meters (250,000 cubic feet) of sludge and solids contaminated with plutonium and americium. Pit 9 was to serve as a demonstration for cleanup technologies that could be applied elsewhere on the site. However, the project has been plagued by significant delays and cost overruns and recent concerns that drilling to retrieve waste samples could cause an explosion or fire. Remediation efforts currently

SIDEBAR 2.5 HISTORICAL WASTE MANAGEMENT PRACTICES IN THE DOE COMPLEX

This April 1962 photograph was taken a few days after rapid melting and rain caused flooding of a pit in what is now the Radioactive Waste Management Complex at the Idaho site. Barrels and boxes containing mixed (radioactive and hazardous) waste can be seen floating in the pit. Source: Idaho National Engineering and Environmental Laboratory.

The Manhattan Project to develop nuclear weapons was a first-of-a-kind engineering effort that produced a variety of "exotic" (by the standards of the day) radioactive and chemical wastes, frequently in very large volumes. During the ensuing Cold War, U.S. (and Soviet) defense efforts were focused on the production of nuclear warheads, and less attention was given to the management and disposal of associated radioactive and chemical wastes, resulting in significant environmental contamination as illustrated by the examples in this chapter.

are on hold awaiting a safety assessment by a team of independent experts.

The remediation of buried waste grounds like the Radioactive Waste Management Complex presents several challenges to DOE and its contractors, including locating and characterizing the buried waste, determining the amount of surrounding contamination, and treating the waste either by in situ or extractive technologies. The problems at this pit provides perhaps a worst-case illustration of the kinds of problems that DOE is likely to face as it tackles other waste burial grounds around the complex.

Burial Grounds at Oak Ridge National Laboratory

The original mission of the Oak Ridge National Laboratory was to produce and chemically separate plutonium, and later to produce isotopes and undertake research on radioactive and hazardous materials. Much of the radioactive and hazardous wastes from these activities is

The reprocessing of spent fuel to recover uranium and plutonium for warheads produced very large volumes of highly radioactive liquid wastes at the Hanford, Savannah River, and Idaho sites, ranging from radioactive or chemically contaminated reactor effluent discharges into groundwater or surface water and soil to high-level waste discharges into the subsurface. The Hanford Site, for example, could not build enough tanks to hold all the waste from reprocessing operations. Consequently, during the 1940s some high-level waste was discharged directly into the ground; and until the 1970s millions of liters of high-level waste supernatant liquids were discharged into the ground through drainage basins and cribs.

One of the guiding philosophies of waste management throughout the DOE complex, especially prior to the 1980s, can perhaps best be characterized as "out of sight, out of mind." Such radioactive and chemical wastes as tritium, chromium, mercury, lubricating oils, solvents, and raw sewage were discharged directly into surface waters, surface drainage basins, or directly into aquifers through injection wells. Solid and liquid radioactive and chemical wastes were also buried in shallow pits and trenches, which are now known by the somewhat euphemistic term "burial grounds." Some of these trenches filled with water during periods of high rainfall, which promoted migration of chemicals and radionuclides into the subsurface.

Many of these waste management practices seem reckless by today's standards, but it is important to recognize that DOE's (and its predecessor agencies) practices were not substantially different from those employed elsewhere in the public and private sectors. In some cases, waste management decisions were made with an incomplete understanding of their consequences. In other cases, waste management practices judged to be appropriate by the standards of the day are now understood to be inadequate in light of our improved understanding of natural processes and our greater sensitivity to environmental quality. Such practices have resulted in a significant legacy of environmental contamination that will take decades and tens to hundreds of billions of dollars to correct.

SIDEBAR 2.6 CONTAMINANT TRAVEL TIMES AT THE RADIOACTIVE WASTE MANAGEMENT COMPLEX

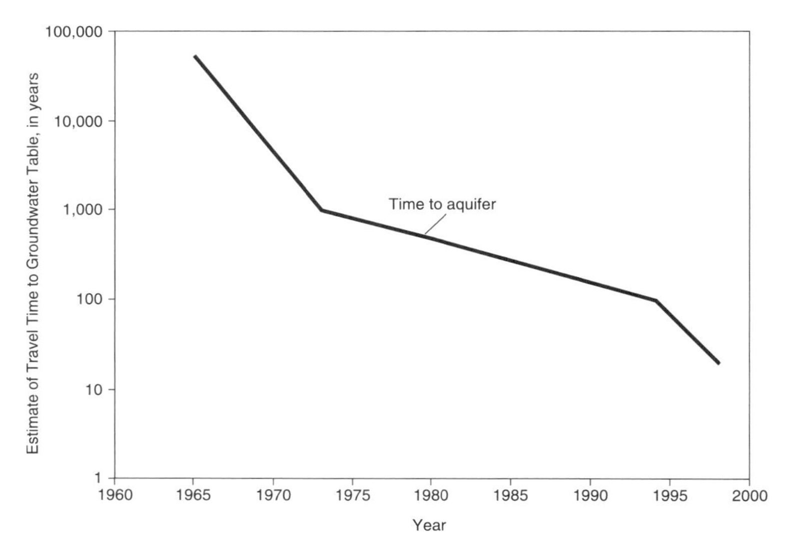

Source: Idaho National Engineering and Environmental Laboratory.

Low levels of plutonium and other contaminants were detected recently in groundwater monitoring wells near the Radioactive Waste Management Complex at the Idaho Site, indicating that contaminants had traveled from the complex, through the unsaturated zone, and into the Snake River plain aquifer. This discovery was unexpected by DOE, since its conceptual models treated the unsaturated zone as a barrier to contaminant migration, and numerical models based on conventional flow and transport theory did not predict this degree of migration.

Travel time from the complex to the underlying Snake River plain aquifer has been the subject of intense debate spanning several decades. Because of site aridity, it was initially assumed that the thick unsaturated zone beneath the complex afforded a high degree of contaminant retardation, but even 40 years ago concerns were raised about the assumption of a long travel time. A National Research Council committee visited the Idaho Site (then the National Reactor Testing Station) and the Hanford Site in the 1960s and prepared a report to the Atomic Energy Commission (NRC, 1966). That committee made the following statement in its report (p. 5):

> **The protection afforded by aridity can lead to overconfidence: at both sites it seemed to be assumed that no water from surface precipitation percolates downward to the water table, whereas there appears to be as yet no conclusive evidence that this is the case, especially during periods of low evapotranspiration and heavier-than-average precipitation, as when winter snows are melted.**

Travel time estimates developed over the last several decades have borne out that committee's concerns. As shown in the figure, travel time estimates have decreased from tens of thousands to a few tens of years. The uncertainty of these estimates is attributed to several factors, including incorrect conceptualizations of the hydrogeologic system, improper simplifying assumptions, incorrect transport parameters, and overlooked transport phenomena.

buried at the site in the Melton Valley Area (DOE, 1996). For example, the Waste Area Grouping 4, which is located about one-half mile southwest of the main plant, is contaminated with strontium-90, tritium, cesium-137, and a small amount of cobalt-60. Significant amounts of tritium have migrated into White Oak Creek, which drains the site (DOE, 1996). About 70 percent of the strontium-90 discharge from this waste area group has been attributed to seepage during waste trench flooding.

There are no cost-effective methods for locating and characterizing these highly concentrated zones of contaminants (known as "hot spots") prior to extraction and treatment. Since waste that must be excavated and moved poses added hazards to workers, most of the buried waste will remain in its current location until more effective technologies become available. Caps and other types of barriers will be used for short-term stabilization and containment, with long-term monitoring to validate the effectiveness of the containment systems. The long-term performance of these containment systems and methods for validating their long-term effectiveness are not well understood.

Soil[4] Contamination

Contamination of surface and near-surface environments is a pervasive problem at all of the major DOE sites. This contamination includes metals, radionuclides, and hazardous chemicals and is the result of poor waste management practices, such as those illustrated below.

Plutonium Contamination at Rocky Flats

As discussed in Sidebar 2.1, the Rocky Flats Environmental Technology Site was responsible for fabrication and component assembly for nuclear weapons. Materials used in these activities included both plutonium and enriched uranium metals and oxides. At present, the Rocky Flats site contains approximately 12.9 metric tons of plutonium and 6.7 metric tons of highly enriched uranium in nuclear weapons parts, materials, process residues, and wastes. Much of the material has been stored in temporary packaging, and about 30,000 liters (~8,000 gallons) of plutonium solutions and 2,700 liters (~710 gallons) of highly enriched uranium acid solutions are being held in tanks that were not designed for long-term storage (DOE, 1996).

Poor storage and disposal practices have resulted in extensive surface and groundwater contamination at the site and on an adjoining property (see Plate 2). The principal types of soil contaminants include americium, plutonium, and uranium. DOE plans several environmental cleanup activities at the site, including removal of contaminant sources, where possible; stabilization, including installation of caps and barriers,

[4]The term "soil" is used here in the engineering sense to include unconsolidated materials in near-surface environments, typically several meters to 10 or so meters in thickness in both saturated and unsaturated states.

FIGURE 2.3 *Plan view of Oak Ridge site and adjacent waterways to Watts Bar Reservoir showing major areas of mercury and cesium contamination. SOURCE: Oak Ridge National Laboratory.*

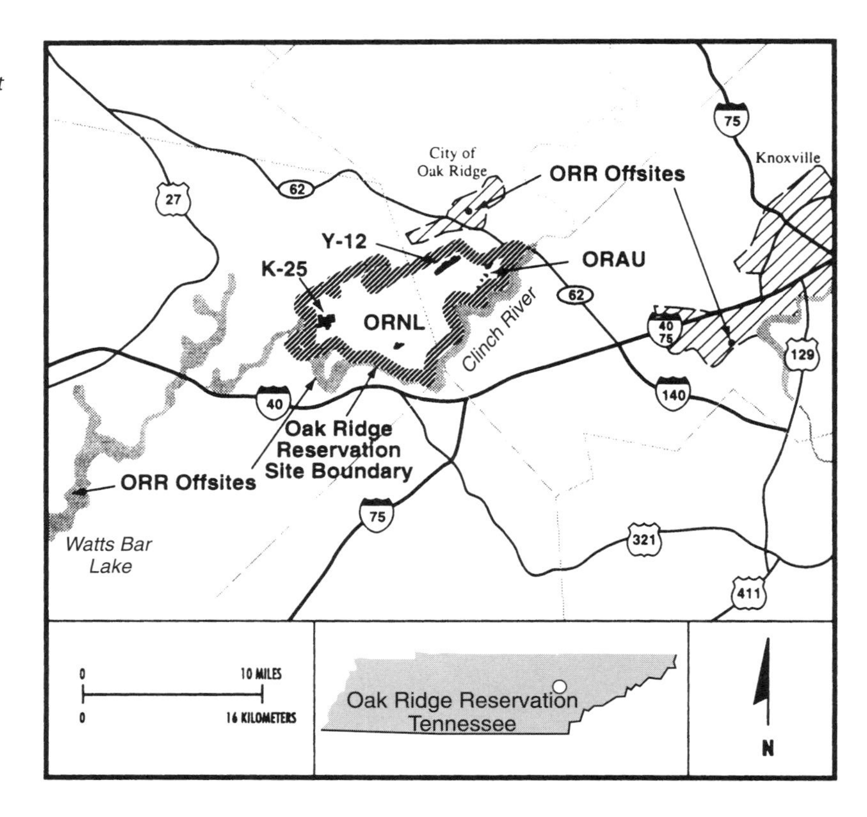

where contamination cannot be removed; and continuous environmental monitoring. DOE has announced plans to complete cleanup of the site by 2006, but even after cleanup is completed there will be a continuing surveillance mission to monitor the remaining contamination (DOE, 1998a).

Mercury and Cesium Contamination at Oak Ridge

Because of poor operational and waste management practices, the streams and rivers on part of the Oak Ridge site have been extensively contaminated with mercury and radioactive cesium. The mercury contamination is from the Y-12 plant, where mercury was used to separate lithium isotopes. DOE estimates that between 108,000 and 212,000 kilograms (~240,000 to 470,000 pounds) of mercury were released into East Fork Poplar Creek between 1953 and 1983 (DOE, 1996). Minor amounts of mercury continue to be released into the creek from secondary sources. The cesium contamination is the result of seepage into streams from old waste storage pits and trenches. These streams drain into the Clinch River, which in turn drains into the Watts Bar Reservoir downstream of the site. The Clinch River and Watts Bar Reservoir comprise about 120 river miles (193 kilometers) and 18,000 hectares

(44,000 acres) and are used for municipal and industrial water supplies, recreation, and residential development (see Figure 2.3 and Plate 5).

Studies by Olsen and others (1992) suggest that about 335 curies of cesium-137 were released into the river system between 1949 and 1986 and that over 300 curies of cesium now reside in the Clinch River and Watts Bar Reservoir sediments. It has been estimated that about 76 metric tons of mercury have accumulated in the sediments of the Watts Bar Reservoir system. Other contaminants found in the river and reservoir system include metals (lead, arsenic, selenium, and chromium), organics (polychlorinated biphenyls and dioxin) and radionuclides (cobalt-60, tritium, and strontium-90).

DOE plans to excavate and dispose of some of the contaminated soils at the Y-12 site. However, there are no plans at present to remediate the river or reservoir, in large part because the contamination is difficult to locate and remediation would be expensive and potentially hazardous to workers, the public, and the environment.

Surface Contamination at Nevada Test Site

There is a significant amount of surface and shallow surface soil contamination that resulted from above-ground and near-surface nuclear detonations, safety shot tests, rocket engine development, and underground nuclear testing at the Nevada Test Site. The primary contaminants include americium, plutonium, depleted uranium, and metals such as lead. The contamination is found on parts of the test site, the Tonopoh Test Range, and the Nellis Air Force Range (see Figure 2.4). The safety shot tests resulted in dispersion of contaminants in excess of 40 picocuries per gram over more than 1,200 hectares (3,000 acres). This contaminated acreage increases to 11,000 hectares (27,000 acres) when atmospheric and near-surface tests are included (DOE, 1996).

When warranted, cleanup of the Soils Sites Area will consist of excavation and disposal elsewhere on the site. Few of these sites have been characterized because of funding constraints.

Contamination in the Unsaturated Zone

The unsaturated zone is that part of the subsurface above the water table. It contains liquid water under less than atmospheric pressures (e.g., water held by capillary and adsorptive forces), but most of the pore spaces in the rock or soil are filled with air. The unsaturated zone exists at all of the major DOE sites, but as shown in Table 2.2 its thickness varies significantly among sites. The unsaturated zone tends to be the thickest at the arid western sites—at Hanford, for example, the unsaturated zone is up to about 90 meters (~300 feet) thick—and thinnest at the more humid eastern sites.

Radionuclide Contamination in the 200 Area at Hanford

The 200 Area is located on what is known as the central plateau of the Hanford Site and covers about 2,400 hectares (6,000 acres; see Plate 3). This area contains chemical processing facilities for extracting uranium and plutonium from irradiated reactor fuel and associated waste storage and facilities. The waste disposal facilities include surface settling basins and underground drainage cribs constructed for disposal of low-activity liquid wastes, as well as solid waste burial pits and trenches. The waste storage facilities include 18 tank farms that contain 177 underground storage tanks containing about 200 million liters (54 million gallons) and about 200 million curies of high-level waste from the separations process. The tanks range in size from about 210,000 liters (55,000 gallons) to about 4.5 million liters (1.2 million gallons) and consist of one or two carbon steel liners surrounded by reinforced

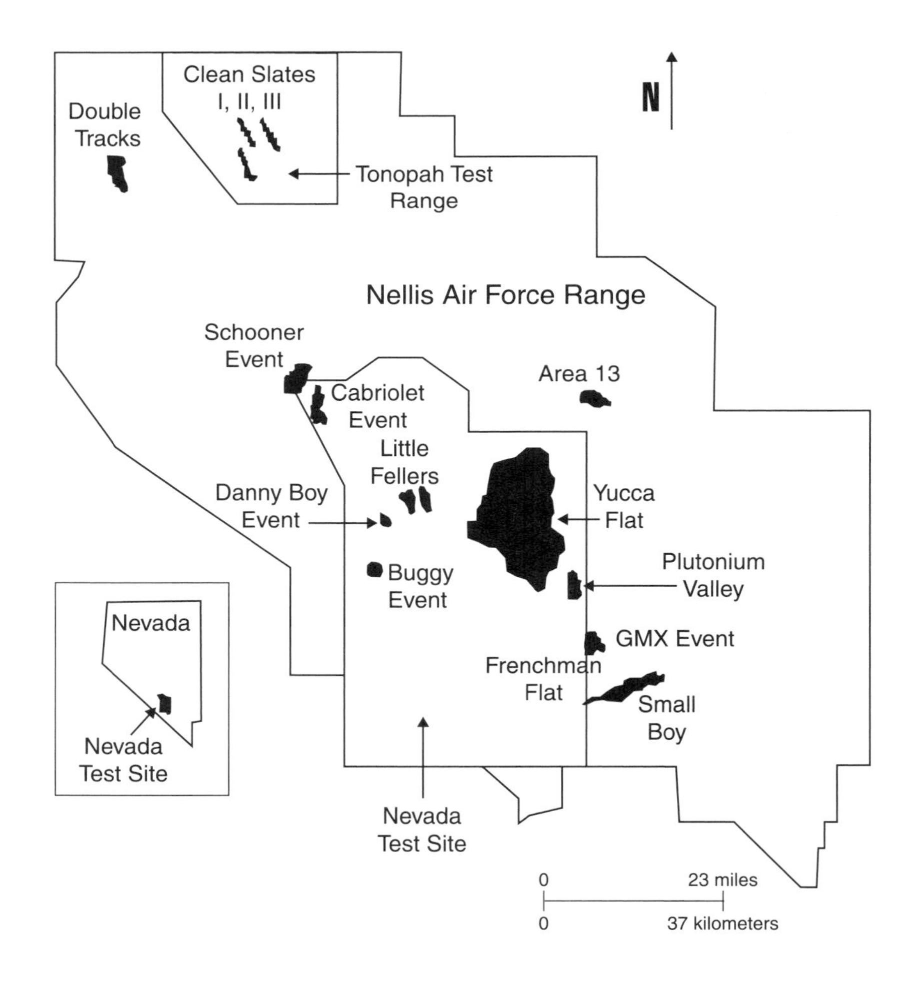

FIGURE 2.4 Plan view of Nevada Test Site showing areas of surface contamination from nuclear testing. SOURCE: Nevada Operations Office.

concrete (DOE, 1996).

DOE estimates that about 1.3 trillion liters (346 billion gallons) of water contaminated with radionuclides were intentionally discharged into the ground through settling ponds and other subsurface drainage structures from chemical processing operations (DOE, 1997a). Additionally, DOE estimates that 67 of the underground storage tanks have leaked at least 3.8 million liters (1 million gallons) of high-level waste into the subsurface. Recent work by Agnew and others (1997), however, suggest that these estimates may be low.

Most of the discharged wastes were supernatant liquids that were produced by gravity-induced settling by allowing the high-level waste to cascade through a series of tanks. These liquids contain such fission products as cesium, strontium, and technetium, as well as short-lived radionuclides like tritium. Later, tank waste evaporators were installed to further reduce waste volumes, and the radionuclide-bearing evaporator sediments were discharged into the soil.

The decisions to dispose of this waste to the soil were based in part on assumptions about the capacity of the unsaturated zone to trap and hold radionuclides through physical and geochemical processes. The unsaturated zone beneath the 200 Area is thick (60 to 90 meters, or 200-300 feet) and contains sand, silt, and gravel above a layer of volcanic rock that was thought to be highly sorptive of radionuclides. Given the small amount of precipitation and high evaporation rates, it was assumed that it would take a long period of time for the contaminants to migrate through the unsaturated zone and into the groundwater (DOE, 1998b).

Technetium-99 well in excess of drinking water standards has been detected in the groundwater beneath the 200 Area, and boreholes have detected possible cesium and strontium at depth beneath several tank farms, most prominently the SX Tank Farm (see Plate 4). This discovery came as a surprise to DOE, because cesium and strontium were assumed to be immobile in the unsaturated zone, and DOE's models of the unsaturated zone predicted that these radionuclides would not migrate significantly. This finding has prompted a reorganization of the cleanup work and a greater effort to integrate science into cleanup activities at Hanford.[5]

---

[5]As a result of this discovery and at the prompting of Congress, DOE created a new organization (Office of River Protection) and the Groundwater/Vadose Zone Integration Project to coordinate the cleanup activities at the Hanford Site. The project will take an integrated approach to solving the groundwater and vadose zone contamination problems to provide a scientific basis for site decisions (DOE, 1998b).

SIDEBAR 2.7 EFFECTS OF SUBSURFACE HETEROGENEITY ON FATE AND TRANSPORT MODELING AND REMEDIATION

Lawrence Livermore National Laboratory, a DOE facility in California, overlies groundwater contaminated with volatile organic chemicals originating from land disposal of chemicals when the site was used as a naval airfield in the 1940s. There are multiple contamination zones corresponding to different disposal locations, consisting primarily of dissolved trichloroethylene and perchloroethylene groundwater contaminant plumes. The western-most plume stretches for over a mile and is of concern because it is migrating slowly toward municipal water supply wells in the city of Livermore. For over 10 years the site has been subject to intensive hydrogeologic investigation and remedial action (Thorpe and others, 1990). As a result, hundreds of monitoring wells have been installed to provide for geologic characterization of the site, monitor the composition and flow of groundwater, and support the design and implementation of remediation technologies.

To more clearly understand the role and effects of geologic heterogeneity on remediation, Tompson and others (1998) used hydraulic conductivity data from 240 of these monitoring wells to construct a statistical distribution depicting the heterogeneous aquifer beneath the site. For a given realization of this distribution, together with various boundary conditions used to reflect remedial (associated with a remedial pumping well) or ambient conditions, groundwater flow paths can be produced using a finite difference flow model.

To illustrate the effects of the fine-scale heterogeneity on contaminant transport and remedial recovery, hypothetical contaminant pulses were released in each model realization to evaluate plausible migration scenarios over 40 years of ambient conditions and then over 200 additional years of remedial pumping from a well located 1,000 meters from the original source. Model runs indicated a wide range of possible outcomes from one realization to the next. When the total pumping time was allowed to run for 200 years, in some cases most of the contaminant mass was recovered from the model domain, whereas in other realizations as little as one-third of the input mass was recovered. This indicates the drastic effect that spatial variability of aquifer materials—the exact distribution of which is never known in precise detail—can have on predictions of contaminant transport. The variation in the results is indicative of the real uncertainty that would be expected for the behavior of a natural system.

Significant uncertainties in understanding of the inventory, distribution, and movement of contaminants in the unsaturated zone exist at Hanford. Further, attempts to model contaminant fate and transport there have met with mixed success. Inaccurate models can have disastrous consequences when they mislead treatment or containment strategies. Therefore, improved models for predicting contaminant migration are needed to evaluate the impact of such releases into the environment. These models must be based on a good understanding of the subsurface features that control contaminant fate and transport (e.g., see Sidebar 2.7), as well as important transport processes.

Metal and Radionuclide Contamination at Idaho

An important mission at the Idaho site was chemical processing of spent fuel from research and naval reactor programs. After chemical processing, the high-level liquid waste was stored in underground tanks. Idaho managers recognized early on that tank storage space would be insufficient, so the site developed a facility to convert the waste into a powdered ceramic, or calcine, that could be more safely handled and stored. Consequently, Idaho was able to avoid the intentional discharge of high-level liquid wastes into the subsurface.

There have nevertheless been several releases of radionuclides and metals from the single tank farm that supported the site's chemical processing facility. An underground waste transfer line was accidentally ruptured by drilling, and up to 13,700 liters (~3,600 gallons) of high-level waste with a total activity of over 32,000 curies was released into the unsaturated zone between 1956 and 1974. In 1972, another leak in the tank farm released about 52,900 liters (~14,000 gallons) with a total activity of about 28,000 curies. The major contaminants include chromium, mercury, cesium, strontium, plutonium, and iodine. Some of this waste is located in a perched water zone beneath the tank farm, but the extent of waste migration is poorly known.

The Idaho site is characterized by a thick unsaturated zone (see Table 2.2), but this zone overlies one of the largest aquifers in the western United States, the Snake River aquifer, which covers an area of about 26,000 square kilometers (10,000 square miles). This aquifer supplies water to most of central Idaho and provides a major source of recharge to the Snake River. Protection of the aquifer and the river is a high priority at the Idaho site and is driving many of the site's remediation decisions. Decisions about remediation of the radionuclide contamination beneath the tank farms is hampered by a lack of information about the distribution of contamination, as well as the physical and chemical characteristics of the unsaturated zone.[6]

Contamination in the Saturated Zone

The saturated zone is defined as that part of the subsurface where pore spaces are filled with water. In unconfined aquifers, the top of the saturated zone defines the groundwater table. The principal saturated zone contamination problem across the DOE complex are contaminated groundwater plumes (i.e., large volumes of groundwater contaminated with dissolved and complexed chemicals, metals, and radionuclides).

---

[6]The committee was told that the least expensive remediation alternative would cost about $600 million and would involve removal of the perched water zone and pump-and-treat remediation of the underlying aquifer.

These plumes have been formed by the injection or migration of waste into moving groundwater and have length scales on the order of kilometers to tens of kilometers, depending on the nature of the source and the rate and direction of groundwater movement.

All of the major DOE sites contain contaminated groundwater plumes, and in some cases these plumes have migrated off site or are discharging into surface waters. The following examples from the Savannah River, Nevada, Hanford, and Idaho sites are illustrative of plume-related problems across the DOE complex.

DNAPL Plumes at Savannah River

The Savannah River Site contains dozens of groundwater plumes containing DNAPLs, metals, and radionuclides, but the DNAPL plume in the Administrative and Materials Manufacturing Area is perhaps most interesting because of its size and location. That area comprises about 140 hectares (350 acres) in the northern portion of the Savannah River Site and is located less than a mile from the site boundary. Currently a research and development center, the area was first established for the manufacture of production reactor components, including target assemblies and fuel rods (Westinghouse Savannah River Co., 1995).

From the 1950s through the early 1980s, contaminated wastewater from fuel and target manufacturing was pumped through an underground line into a settling basin, which had a capacity of about 30 million liters (8 million gallons). The basin overflowed periodically into a natural seepage area and a shallow depression known as Lost Lake and released approximately 1.6 million kilograms (3.5 million pounds) of solvents (principally trichloroethylene and tetrachloroethylene) and heavy metals to the environment. DOE believes that most of the heavy metals were trapped in the soil and about half of the solvents evaporated, while the remainder migrated downward from the seepage areas into the groundwater (Westinghouse Savannah River Co., 1995). In this part of the site the groundwater moves at rates ranging from a few centimeters to about 90 meters per year.

DOE has installed some 400 monitoring wells since 1981 to track the spread of contamination, and based on these monitoring data and modeling studies, scientists at the Savannah River Technology Center have created a three-dimensional representation of the plume. DOE has installed a pump-and-treat system at the downstream toe of the plume to halt its further spread. DOE has been unable to locate or remove the DNAPL sources that are feeding this plume or to apply effective remediation technologies to the plume itself; it therefore faces the prospect of long-term institutional management of this contamination, including pump-and-treat remediation.

Radionuclide Contamination at the Nevada Test Site

Over 925 nuclear tests were conducted at the Nevada Test Site between 1951 and 1992 and resulted in the emplacement into the subsurface of several hundred million curies of radioactivity, including significant quantities of tritium, plutonium, and fission products (see Table 2.4). Many of these tests were conducted at or below the groundwater table. Nevada officials contend that the site contains more contaminated media than any other site in the DOE complex (Walker and Liebendorfer, 1998). DOE notes in *Paths to Closure* (DOE, 1998a, p. E-56) that it has no plans to remediate the subsurface in and around the underground tests because "cost-effective remediation technologies have not yet been demonstrated."

TABLE 2.4 Isotope Inventories from Underground Testing at the Nevada Test Site

Location	Isotope	Inventory (10^6 curies) (Numbers are rounded)
Pahute Mesa[a]	Tritium	69.9
	Cesium-137	1.95
	Strontium-90	1.56
	Krypton-85	0.13
	Plutonium-241	0.09
	Samarium-151	0.07
	Europium-152	0.03
	Plutonium-239	0.02
	Europium-154	0.02
	Others (34 isotopes)	0.05
	Total Pahute Mesa	73.8
Non-Pahute Mesa	Tritium	30.7
	Potassium-40	24.7
	Cesium-137	1.48
	Strontium-90	1.19
	Plutonium-241	0.10
	Krypton-85	0.09
	Europium-152	0.06
	Samarium-151	0.05
	Europium-154	0.05
	Plutonium-238	0.03
	Plutonium-239	0.01
	Others (32 isotopes)	0.04
	Total Non-Pahute Mesa	58.5

[a]See Figure 2.5 for locations.

SOURCE: Presentation to the committee by Robert Bangerter, DOE-Nevada Operations Office, December 15, 1998.

SIDEBAR 2.8 PLUTONIUM MIGRATION AT NEVADA TEST SITE?

A potentially significant example of the deficiency in understanding subsurface radionuclide transport processes was provided by Karsting and others (1999), who reported that they had detected plutonium in groundwater at the Nevada Test Site. The plutonium was detected in water collected from monitoring wells on Pahute Mesa, near the northwestern border of the test site (Figure 2.5). The plutonium was apparently being carried on colloids. The origin of the colloids and the plutonium geochemistry is still uncertain.

Karsting and others were able to trace the plutonium to the Benham Test, which was detonated in 1968 in zeolitized bedded tuff at a depth below the surface of about 1,400 meters. This test is located about 1.3 km laterally and up to 600 meters below the monitoring wells. The origin of the plutonium was identified from its $^{240}Pu/^{239}Pu$ isotopic ratio, which is distinctive for each underground test. The plutonium ratio is recorded in the melt-glass collected from the underground test cavities. No evidence was found for migration of plutonium from other nearby tests.

The suggested transport of plutonium at the test site has potentially significant implications for DOE's plans to passively manage contaminants there, especially if plutonium transport proves to be more pervasive than is currently recognized. This discovery also has potentially significant implications for the underground disposal of nuclear waste. Conventional wisdom suggests that plutonium is relatively immobile in oxidizing subsurface environments like at the test site and has strong sorbing tendencies. Indeed, underground tests at the test site were believed to demonstrate the effective fixation of plutonium in subsurface environments. The work by Karsting and others has demonstrated that the conceptual models for plutonium migration are incomplete; it also suggests that additional basic research on the geochemical behavior of plutonium is required.

Tritium is very mobile in groundwater, and large plumes of tritium have been detected from many of the underground tests. It has long been argued that most other radionuclides, and especially plutonium, are relatively immobile due to their low solubilities in groundwater and strong sorption onto mineral surfaces. As discussed in Sidebar 2.8, however, recently published work challenges this conventional view.

Mixed Contaminant Plumes at Test Area North

Test Area North at the Idaho National Engineering and Environmental Laboratory covers about 50 hectares (125 acres) in the northern part of the site and was used to support the Aircraft Nuclear Propulsion Program between 1954 and 1961. From 1960 through the 1970s, the area housed the Loss-of-Fluid Test Facility, which was used for reactor safety testing and behavior studies. The primary source of the contaminated groundwater plume is the Technical Support Facility injection well, which was used from 1953 to 1972 to inject liquid wastes directly into the Snake River plain aquifer. The contaminants included raw sewage, trichloroethylene, tritium, strontium-90, and cesium-137.

Although the source area for this plume—the injection well—is known, the source term is not. Moreover, the subsurface in this region consists of highly fractured rock, which makes it difficult to locate and characterize the contamination. Characterization of the extent of contamination began in 1988, and recent data suggest that most of the contamination probably occurred as entrained sludge in two major fracture zones (see Figure 2.6).

Contaminant Plumes at the Hanford Site

DOE estimates that groundwater under more than 220 square kilometers (85 square miles) of the Hanford Site is contaminated above current standards, mostly from operations in the 100 and 200 Areas (Plate 3). The 100 Area is located on about 6,900 hectares (17,000 acres) in the northern section of the Hanford site and contains nine production reactors and several waste burial sites (DOE, 1996). The main sources of subsurface contamination in the 100 Area are from radionuclide (mainly tritium) contaminated reactor cooling water and metal and DNAPL contaminants from operations and disposal. Contamination in the 200 Area was discussed in the section on the unsaturated zone earlier in this chapter.

Disposal of supernatant liquids into the ground and leaks from the high-level waste tanks have produced significant contamination of the saturated zone in the 200 Area (Gephart and Lundgren, 1998). Groundwater plumes of the following contaminants exist at levels exceeding current drinking water standards at the 200 Area: tritium, strontium-90, technetium-99, iodine-129, carbon tetrachloride, chromium, and uranium. The plumes are flowing northeast toward the Columbia River at several tens of meters per year (see Figure 2.7).

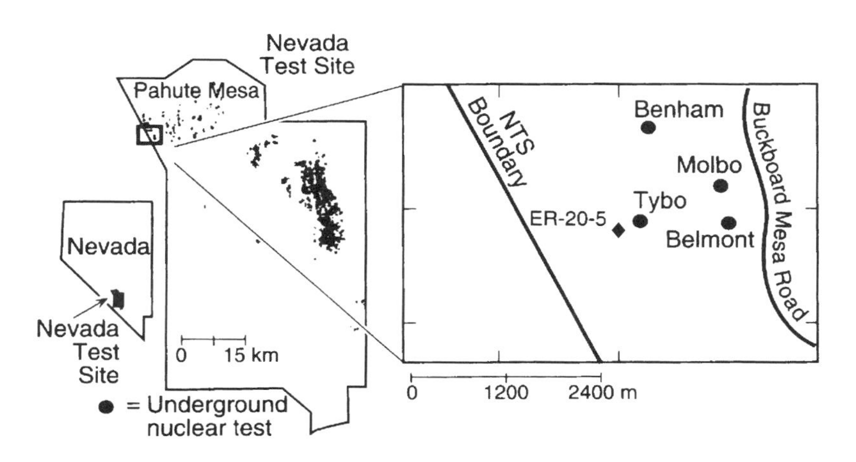

FIGURE 2.5 Plan view of Pahute Mesa with location of the Benham Test and groundwater collection well cluster ER-20-5. SOURCE: Karsting and others (1999).

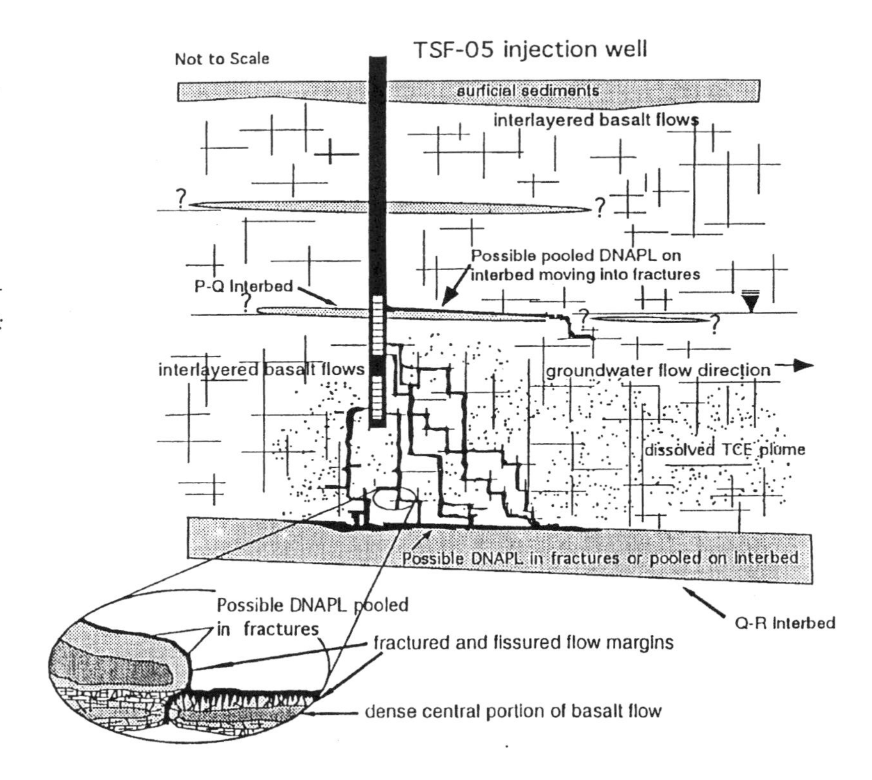

FIGURE 2.6 Conceptual model for subsurface contamination at Test Area North at the Idaho Site. Dense non-aqueous phase liquids (DNAPLs) may be entrained in fractures and perched on dense basalt flows and sedimentary interbeds. SOURCE: Idaho Engineering and Environmental Laboratory.

DOE has established an extensive network of monitoring wells to track the movement of the groundwater plumes, but very little remediation work is being done at present. DOE has established a groundwater extraction well network to intercept a chromium plume in the 100 Area. The chromium is extracted using ion exchange and the treated water is returned to the aquifer. Pump-and-treat systems also have been established in the 200 Area to contain the highest concentrations of a uranium and technetium-99 plume and a carbon tetrachloride plume (DOE, 1998b).

DOE has a very poor understanding of the source areas, amounts, and timing of contaminant discharges into the subsurface at Hanford. DOE is beginning to support "forensic" investigations of past waste releases to the subsurface (e.g., Agnew and others, 1997), but additional work will be needed to improve the knowledge of the extent and magnitude of subsurface contamination at the Hanford site. Improvements in understanding and modeling fate and transport processes in the subsurface is also needed to provide long-term predictive capabilities.

Conclusions

The examples provided in this chapter illustrate that subsurface contamination is an enormously difficult cleanup problem as well as a significant challenge to science. Much of the subsurface contamination at DOE sites is poorly characterized and widely dispersed in the environment, making it very expensive or technically impractical to treat effec-

FIGURE 2.7 Plan view showing the fast spread of tritium plumes from the 200 East Area at the Hanford Site to the Columbia River. SOURCE: Richland Operations Office.

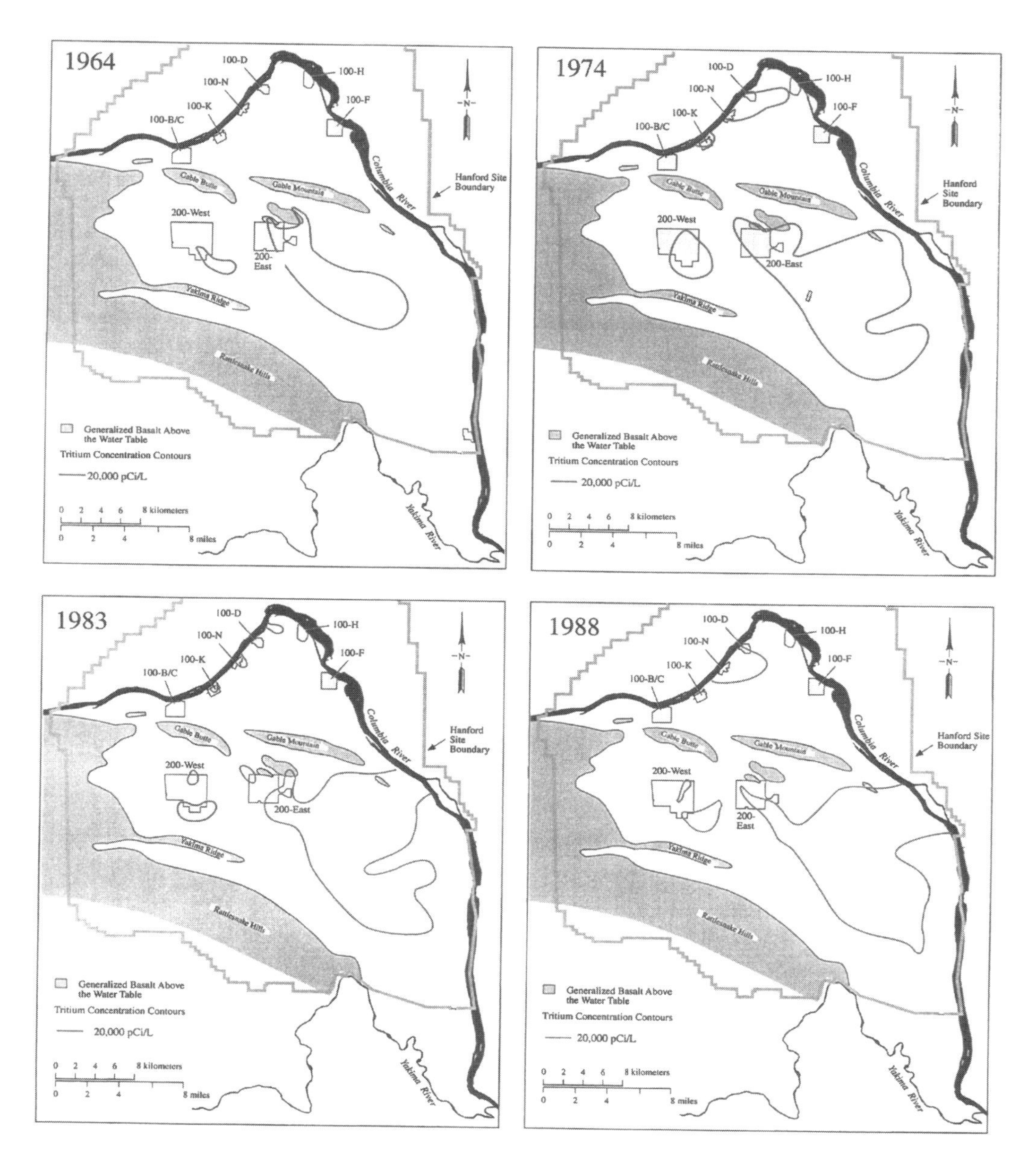

tively with current technologies. Moreover, the contamination that cannot be removed or effectively isolated from the environment will require long-term management, which represents a potentially large future mortgage for the nation.

SIDEBAR 2.9 BASIC SCIENCE CAN IMPROVE ENVIRONMENTAL MANAGEMENT

Basic scientific research can provide several benefits to waste management efforts if it is properly focused on difficult cleanup problems (see Chapter 6). Basic research can produce new scientific knowledge and engineering tools to improve the effectiveness of cleanup efforts, lower cleanup costs, reduce risks to worker and public health, and improve environmental quality. Equally important, basic research can help improve current waste management practices and thereby reduce the likelihood of future environmental insults. Scientific studies in the 200 Area at Hanford provide a simple yet compelling illustration of the potential benefits for environmental management.

The 200 Area is comprised of two major operating zones (200 East and 200 West) that contain a variety of waste disposal and waste storage facilities (see Plate 3). These facilities, which include drainage cribs, settling basins, and underground tanks, are major contributors to the site's groundwater contamination. As discussed elsewhere in this chapter, groundwater contaminant plumes have formed beneath both areas, but the plumes originating from the 200 East Area are significantly larger in size, extending some 15 kilometers (9 miles) to the Columbia River (see Figure 2.7).

Basic geological research conducted at Hanford (see Reidel and others [1992] and DOE [1998b] for a summary of the Hanford geology) suggests that plume size is controlled to a large extent by the physical and chemical properties of the geological formations underlying the 200 Area. The 200 East Area is underlain by the Hanford Formation, which is comprised of permeable sands and gravels that provide relatively direct pathways to the groundwater some 100 meters below the surface. The 200 West Area, on the other hand, is underlain by the Ringold Formation, which consists of less permeable sands, gravels, and clays that provide a barrier to widespread contaminant migration.

These findings provide a compelling demonstration that "geology counts" in waste management and site remediation, and that locating disposal facilities must take account of subsurface properties as part of a defense-in-depth waste containment strategy.[1] DOE is constructing and operating several facilities in the 200 Area to dispose of a variety of cleanup and defense wastes. It recently sited a large land disposal facility (the Environmental Restoration Disposal Facility) in 200 West to manage certain types of chemically and radioactively hazardous cleanup wastes from other parts of the Hanford Site. At least two other disposal facilities have been constructed or are planned for the 200 East Area: the Naval Reactor Disposal Facility, which contains nuclear reactors from decommissioned U.S. Navy submarines, and the planned Immobilized Low-Activity Waste Disposal Facility, which will take low-activity waste generated during processing of high-level waste from the Hanford tanks. If the past is a guide to the future, the disposal facilities in the 200 East Area may create new site contamination problems that will require additional remediation efforts.

[1]A defense-in-depth waste containment strategy uses multiple artificial or natural barriers to improve the long-term performance of the containment system.

The committee believes that this future mortgage could be reduced significantly through the development of new and improved technologies to locate, remove or contain, and monitor subsurface contamination at DOE sites. However, the development of such technologies will require advances in basic understanding of the complex natural systems at DOE sites and also in understanding the nature of contaminant "insults" to those systems. The report of the NRC Committee on Building an Effective Environmental Management Science Program (NRC, 1997b, p. 22) concluded that "new technologies are required to deal with EM's most difficult problems, and new technologies require new science." The present committee agrees with this statement and notes that, given the long-term nature of the cleanup mission and its projected cost (see Chapter 1), DOE has necessary cause and time to do the required basic research to support the development of these needed technologies (see Sidebar 2.9).

Color Plates

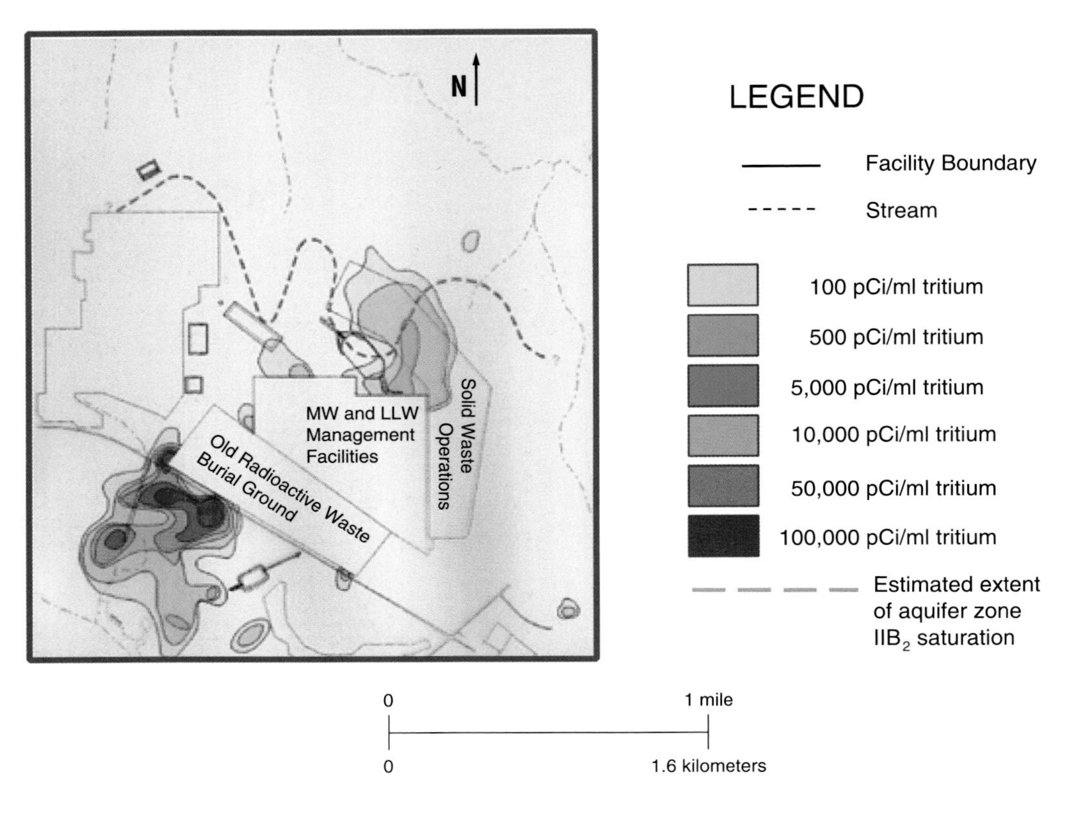
N
LEGEND
Facility Boundary
Stream
100 pCi/ml tritium
500 pCi/ml tritium
5,000 pCi/ml tritium
10,000 pCi/ml tritium
50,000 pCi/ml tritium
100,000 pCi/ml tritium
Estimated extent of aquifer zone IIB$_2$ saturation
MW and LLW Management Facilities
Solid Waste Operations
Old Radioactive Waste Burial Ground
0
1 mile
0
1.6 kilometers

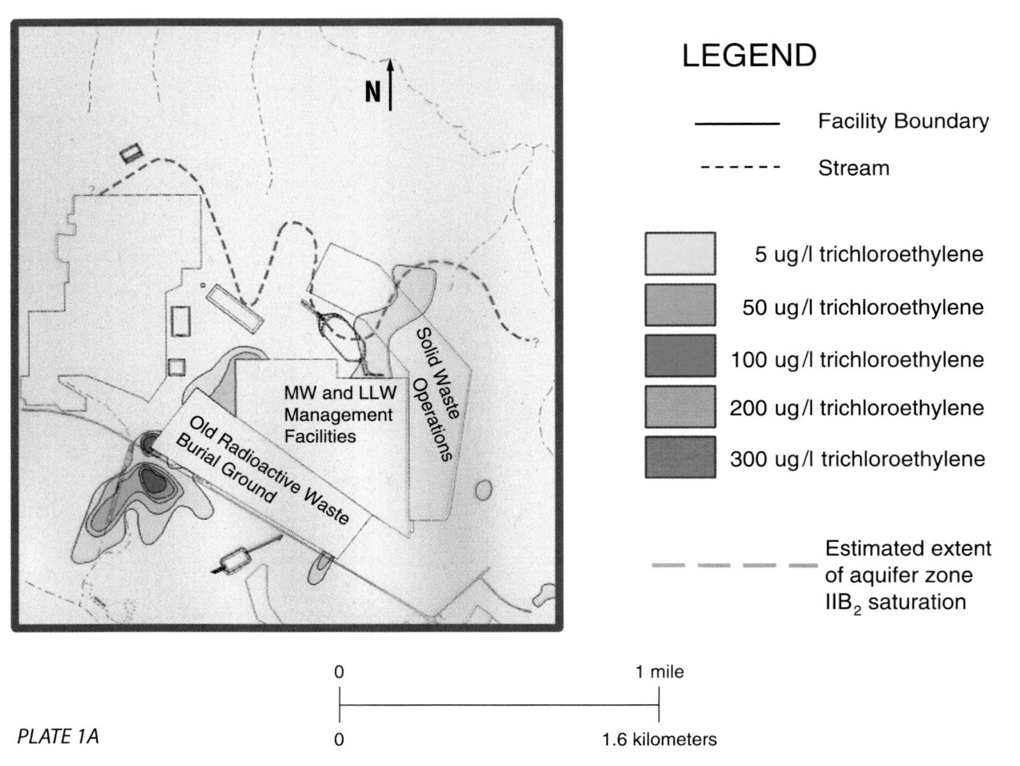
N
LEGEND
Facility Boundary
Stream
5 ug/l trichloroethylene
50 ug/l trichloroethylene
100 ug/l trichloroethylene
200 ug/l trichloroethylene
300 ug/l trichloroethylene
Estimated extent of aquifer zone IIB$_2$ saturation
MW and LLW Management Facilities
Solid Waste Operations
Old Radioactive Waste Burial Ground
0
1 mile
0
1.6 kilometers

PLATE 1A

PLATE 1A Plan view of the Waste Burial Ground Complex at the Savannah River Site and associated tritium (top) and trichloroethylene (bottom) groundwater plumes.

PLATE1B Schematic cross-section of the surface barrier, or geosynthetic cap, and photo of a cap that is being constructed over the old waste burial ground. SOURCE: Savannah River Site.

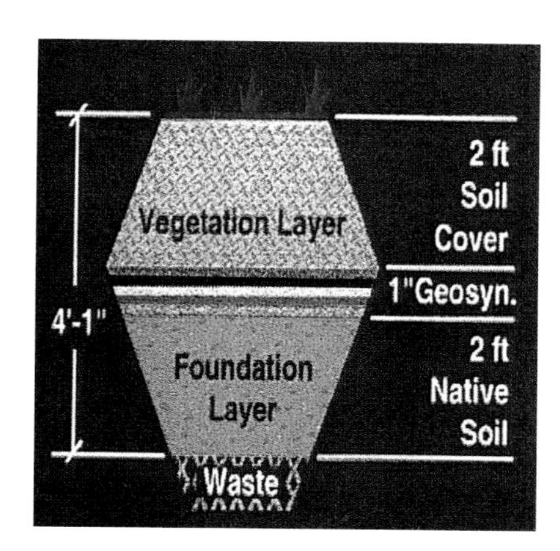

PLATE1B

PLATE 2 *Plan view of Rocky Flats Environmental Technology Site showing areas of plutonium soil contamination.*
SOURCE: Rocky Flats Environmental Technology Site.

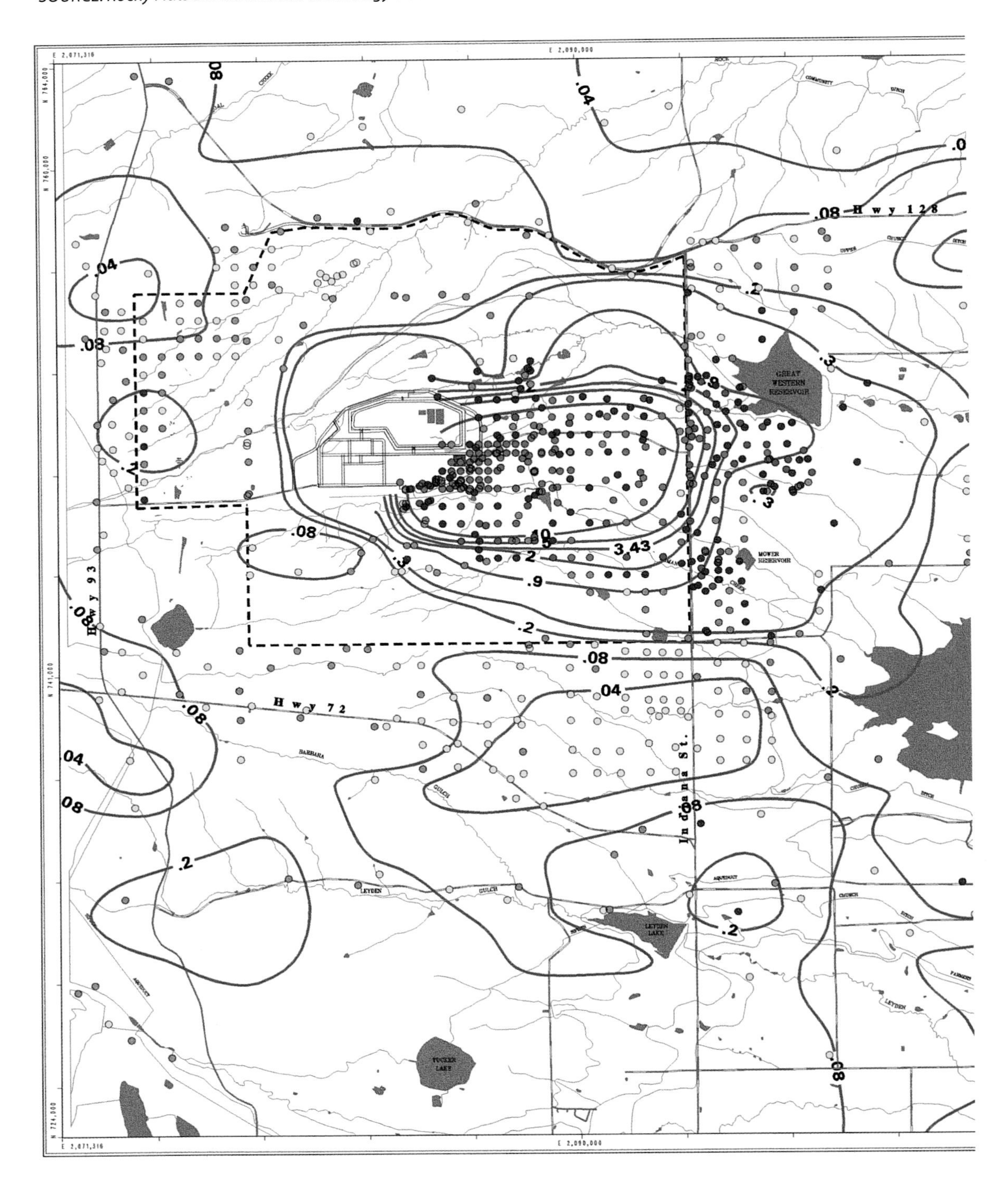

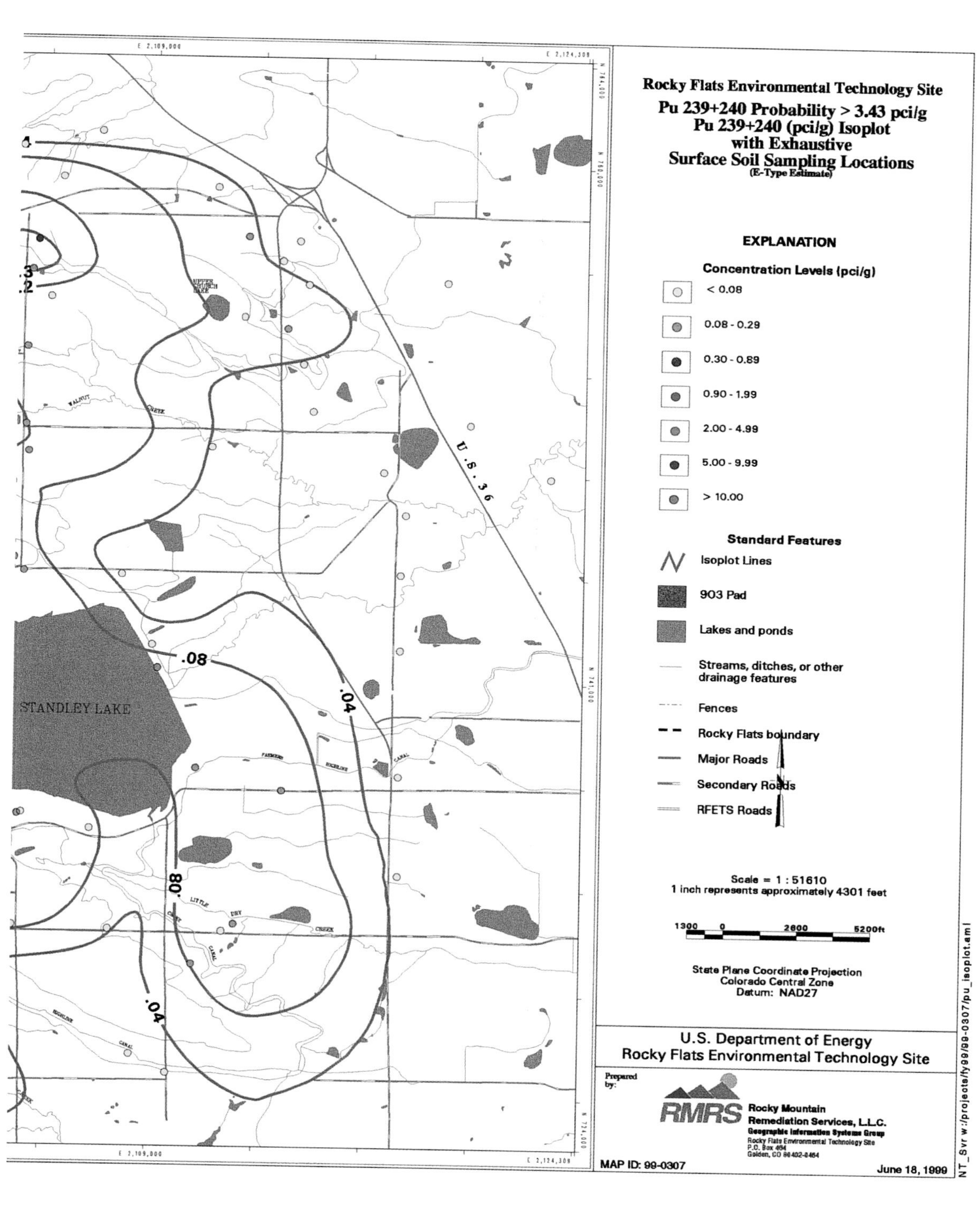
Rocky Flats Environmental Technology Site
Pu 239+240 Probability > 3.43 pci/g
Pu 239+240 (pci/g) Isoplot
with Exhaustive
Surface Soil Sampling Locations
(E-Type Estimate)
EXPLANATION
Concentration Levels (pci/g)
< 0.08
0.08 - 0.29
0.30 - 0.89
0.90 - 1.99
2.00 - 4.99
5.00 - 9.99
> 10.00
Standard Features
Isoplot Lines
903 Pad
Lakes and ponds
Streams, ditches, or other drainage features
Fences
Rocky Flats boundary
Major Roads
Secondary Roads
RFETS Roads
Scale = 1 : 51610
1 inch represents approximately 4301 feet
1300 0 2600 5200ft
State Plane Coordinate Projection
Colorado Central Zone
Datum: NAD27
U.S. Department of Energy
Rocky Flats Environmental Technology Site
Prepared by:
RMRS
Rocky Mountain Remediation Services, L.L.C.
Geographic Information Systems Group
MAP ID: 99-0307
June 18, 1999
NT_Svr w:/projects/fy99/99-0307/pu_isoplot.aml
STANDLEY LAKE
U.S. 36
.08
.04
.3
.2
E 2,109,000
E 2,124,309

PLATE 3 Plan view of Hanford Site showing locations of features in the 100 and 200 Areas and boundaries of major groundwater plumes. The 100 Area is located along the Columbia River and contains the site's production reactors. The 200 Area, which is located in the central part of the site, contains waste management facilities. A. Radionuclide contaminant plumes. B. Hazardous chemical and nitrate plumes. (DWS = drinking water standards.) SOURCE: Richland Operations Office.

PLATE 3A

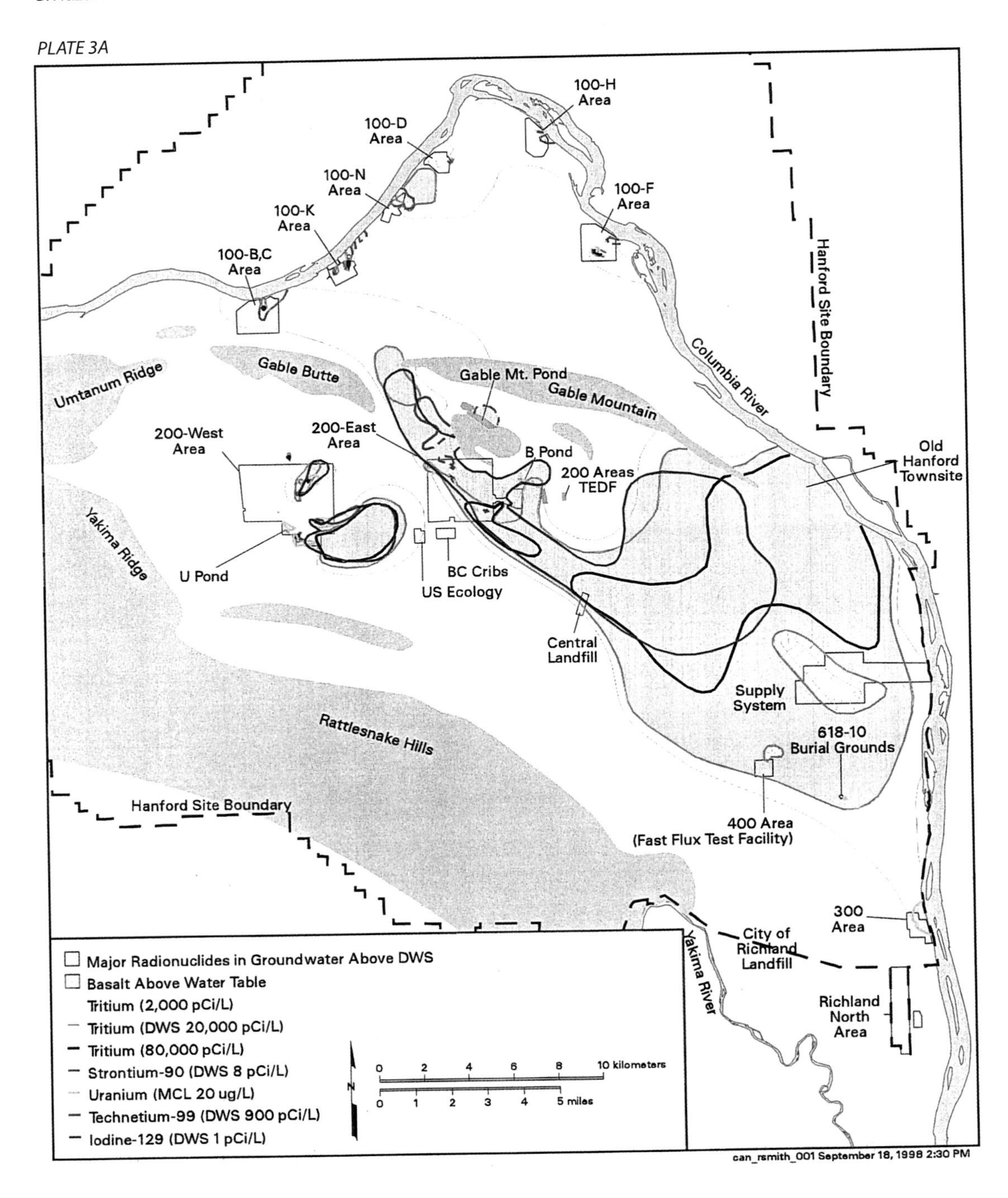

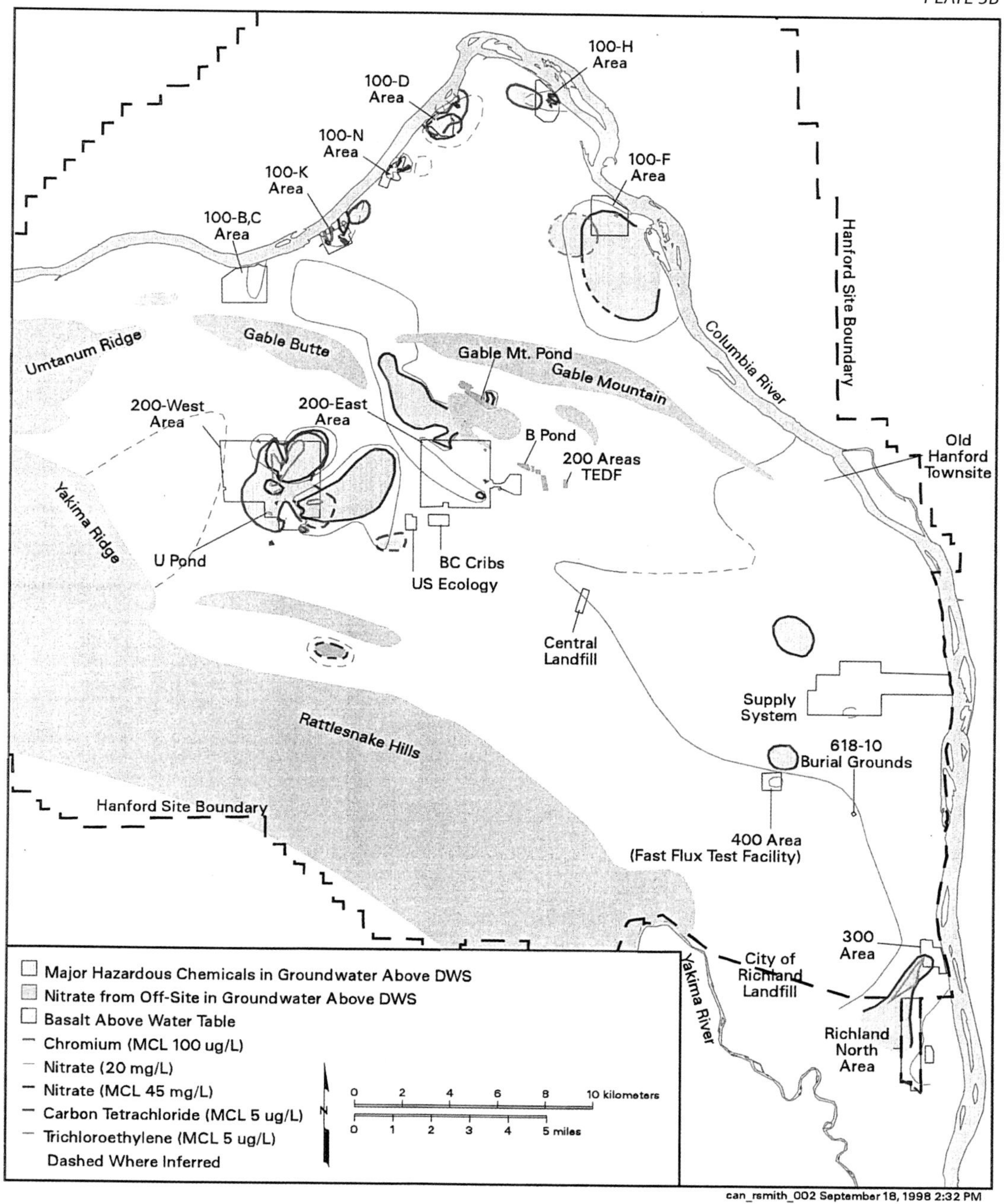
100-H Area
100-D Area
100-N Area
100-K Area
100-B,C Area
100-F Area
Hanford Site Boundary
Columbia River
Umtanum Ridge
Gable Butte
Gable Mt. Pond
Gable Mountain
200-West Area
200-East Area
B Pond
200 Areas TEDF
Old Hanford Townsite
Yakima Ridge
U Pond
BC Cribs
US Ecology
Central Landfill
Supply System
Rattlesnake Hills
618-10 Burial Grounds
Hanford Site Boundary
400 Area (Fast Flux Test Facility)
300 Area
City of Richland Landfill
Yakima River
Richland North Area
Major Hazardous Chemicals in Groundwater Above DWS
Nitrate from Off-Site in Groundwater Above DWS
Basalt Above Water Table
Chromium (MCL 100 ug/L)
Nitrate (20 mg/L)
Nitrate (MCL 45 mg/L)
Carbon Tetrachloride (MCL 5 ug/L)
Trichloroethylene (MCL 5 ug/L)
Dashed Where Inferred
N
0 2 4 6 8 10 kilometers
0 1 2 3 4 5 miles
can_rsmith_002 September 18, 1998 2:32 PM

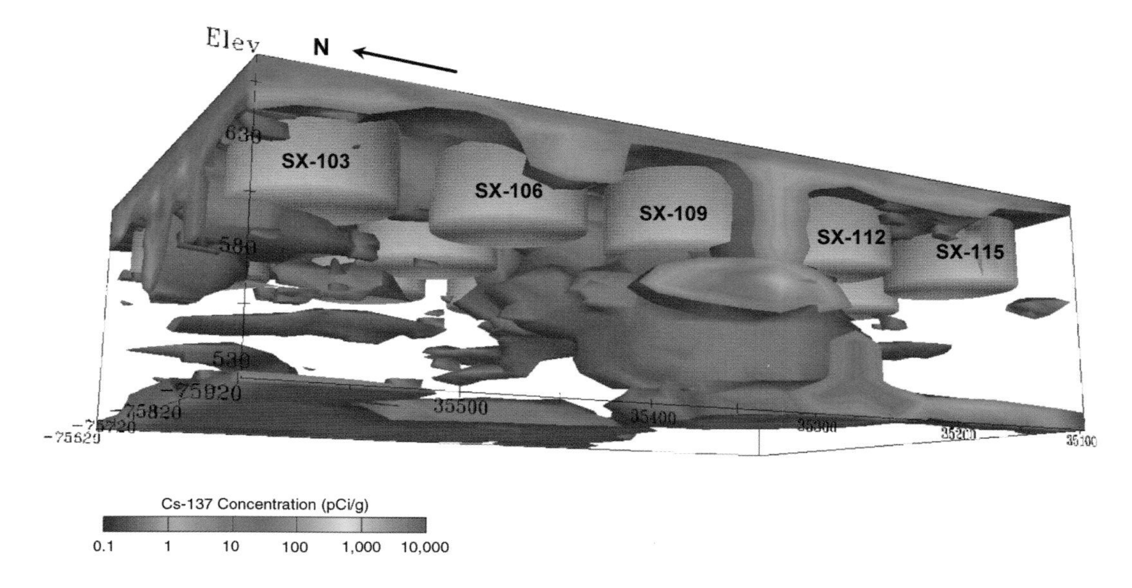

PLATE 4 Oblique view (looking from below the tanks toward the surface) showing cesium-137 contamination beneath the SX tank farm in 200 East Area at Hanford. Elevation is shown in feet above mean sea level, and the visible tanks are numbered with the prefix SX-. SOURCE: Richland Operations Office.

PLATE 5 Photo of mercury found in soils at the Y-12 Plant at the Oak Ridge Site. SOURCE: Oak Ridge National Laboratory.

3

Assessment of the EM Science Program Portfolio

The Environmental Management (EM) Science Program[1] has been in existence for about four years and has completed four proposal competitions.[2] The program has supported research projects relevant to many aspects of DOE's cleanup program, for example, research on subsurface contamination, high-level waste, and deactivation and decommissioning. In its 1998 report to Congress (DOE, 1998g), DOE identified 82 EM Science Program projects with a total investment of approximately $70 million[3] that address the remedial action problem area, which focuses on the cleanup of soil, surface water, and groundwater at sites where contaminants or contaminated materials have been spilled, dumped, disposed, or abandoned (DOE, 1998a, p. 2-9).

The first two proposal calls did not provide detailed descriptions of DOE's cleanup problems, and the proposal review process (see Appendix A) focused first and foremost on identifying scientifically meritorious projects for funding. Relevance to DOE's problems was considered only for those projects that were deemed to be of high scientific quality. Thus, as this committee began to address its task statement to provide advice on a subsurface research agenda (see Chapter 1), it asked itself the following two questions, which provide a focus for the current chapter of this report:

---

[1]As discussed in Chapter 1, the current program was established by Congress in fiscal year 1996. Previously, the Office of Science and the Office of Environmental Management (EM) jointly managed a one-year pilot project that awarded about $5 million in 3-year grants for research on EM-related projects.

[2]The four completed competitions were held in fiscal years 1996, 1997, 1998, and 1999. The 1999 competition was completed while this report was in review.

[3]Many of the awards are being funded over multiple years and are therefore subject to future congressional appropriations.

1. To what extent does the EM Science Program research portfolio for fiscal years 1996 and 1997 address DOE's significant subsurface contamination problems?
2. In light of these current investments, are there any particular subsurface problems that should be emphasized in future proposal calls?

The committee reviewed all projects awarded funding during the first two proposal competitions (in fiscal years 1996 and 1997) and attempted to assess the extent to which these projects addressed the cleanup problems identified in Chapter 2. The word "attempted" is used advisedly, because these projects were still in progress at the time of the committee's review and therefore the research results were incomplete. Moreover, the committee did not review research results for scientific merit in the way that one would review papers submitted to refereed journals, so it cannot comment on the quality of the work resulting from these projects. The committee's assessment is based on a review of project titles, principal investigator experience and affiliations, project abstracts as provided in DOE's 1998 report to Congress (DOE, 1998g), and on a review of progress reports provided by the principal investigators, which were published in the proceedings volume of the Environmental Management Science Program Workshop (DOE, 1998c). This

TABLE 3.1 Summary of the EM Science Program Portfolio for Fiscal Years 1996 and 1997 and Pilot Projects Funded in Fiscal Year 1995

Category	Projects Funded[a]	Research Focus				Methodology			Number of Multiple Investigator Projects
		Metals	Organics	Radionuclides	Site[b]	Field	Laboratory	Modeling/Theory	
Identify	30	1	12	3	16	23	12	19	21
Contain	6	3	1	4	NA	1	4	2	3
Remediate	37	14	24	10	NA	5	35	6	17
Remove	7	6	1	0	NA	1	7	0	1
Validate	9	1	4	2	3	4	5	3	5
Other	16	6	8	7	NA	4	13	2	6

[a]This column sums to 105 projects, because some projects were included in more than one category. There are 91 separate projects represented by the data in this table.

[b]Projects that focused on characterization of the site rather than on specific contaminants.

SOURCE: DOE (1998c,g).

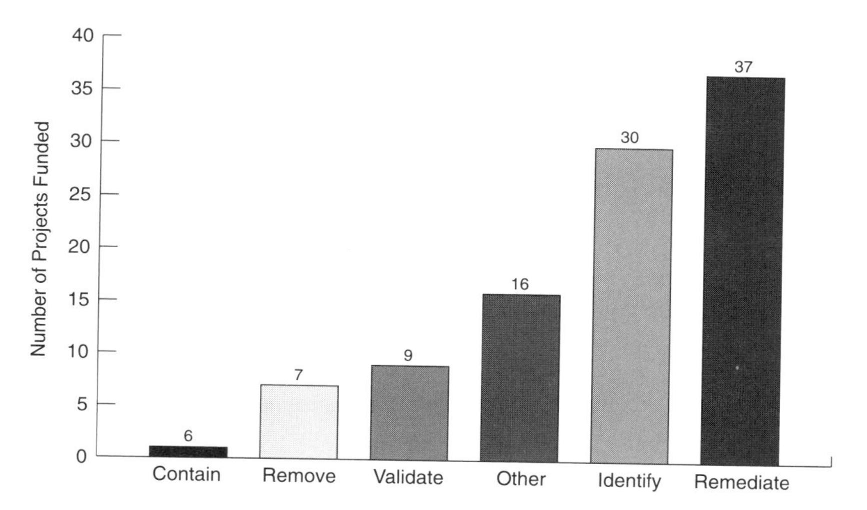

FIGURE 3.1 Distribution of subsurface research projects in the EM Science Program portfolio for fiscal years 1996 and 1997. The numbers in the graph are the number of projects funded in each topical area.

workshop was held in Chicago, Illinois on July 27-30, 1998. These analyses are summarized in Table 3.1 and Figure 3.1.

The committee spent a considerable amount of time during its first two meetings discussing the merits of various organizing schemes for this assessment and eventually adopted a slightly modified form of an approach that is used by DOE's Subsurface Contaminants Focus Area[4] to organize its technology development programs (see Figure 3.2). This organizational scheme comprises a five-point technical strategy that is based on what the focus area refers to as "the accepted process for the remediation of contaminated sites" (DOE, 1997b). This scheme considers the generic processes that must be employed to remediate a site (e.g., locate the waste, treat the waste, validate the treatment process) without reference to the specific technologies that will be employed to accomplish these processes. The committee adopted the focus area's

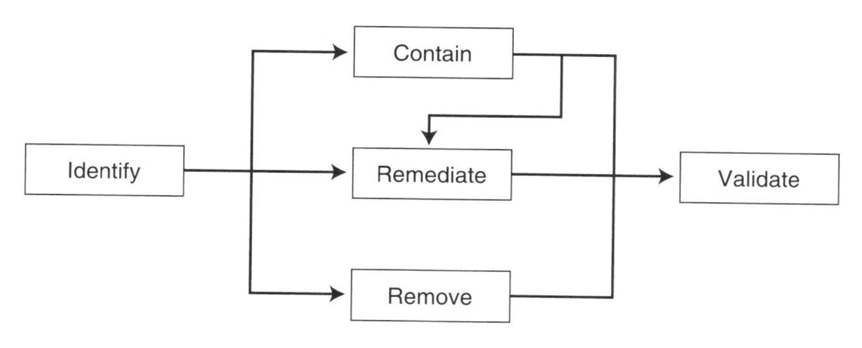

FIGURE 3.2. Flow chart for remediation of subsurface contamination (DOE, 1998d, p. 4).

[4]The Subsurface Contaminants Focus Area is part of the Office of Science and Technology, which is responsible for developing technologies for cleanup of the DOE complex. The EM Science Program is also part of this office.

function names, slightly modified some of the function descriptions, and added an additional category ("Other") to its analysis to contain those projects that do not fit readily into one of the focus area's categories.

The resulting organizing scheme used for the committee's assessment is shown below:

- *Identify*—Locate and quantify suburface contamination.
- *Contain*—Contain or stabilize mobile contaminants and locally elevated contaminant concentrations (i.e., contaminant hot spots) in situ.
- *Remediate*—Treat to reduce mobility or destroy mobile contaminants in situ.
- *Remove*—Extract contaminant hot spots that are not amenable to in situ treatment.
- *Validate*—Verify conceptual models and the performance of remediation processes or strategies.
- *Other*—Projects that address subsurface contamination problems, but do not fit into one of the preceding categories.

The committee adopted this scheme for organizing its assessment mainly for convenience, but also because this scheme could provide a direct linkage between basic research in the EM Science Program and applied technology development in the Subsurface Contaminants Focus Area. As will be discussed in Chapter 6, moving the results of basic research from the EM Science Program into application at the sites is a major challenge confronting DOE. The committee hopes this organizing scheme will provide a useful mechanism for identifying potentially fruitful application paths for EM Science Program-sponsored research.

A summary of the committee's assessment of the current program portfolio is provided in the following sections. A concluding section provides a brief discussion of the two questions posed at the beginning of this chapter.

Identify

The radioactive and hazardous subsurface contaminants of concern at DOE sites (see Chapter 2) have entered the soil and groundwater through accidental spills, poor waste management practices, and failure of storage and containment systems. Even in cases where the points of contaminant entry into the subsurface are known, information on timing of entry and contaminant quantities may be lacking. Once introduced into the subsurface, the contaminants are subject to a number of physical and chemical processes or biological degradation. Subsurface heterogeneities may make it difficult to predict contaminant movement away from release sites. Successful remediation of contaminated subsurface sites depends first and foremost on the ability to locate and quantify the nature and extent of contamination, the focus of this category.

The committee found 30 projects relevant to the "Identify" category in the portfolio (see Table 3.1). These projects encompass a wide range of topics and approaches, but in general focus on the following: (1) location and spatial distribution of contaminants in saturated and unsaturated environments; (2) methods to estimate quantitatively the extent of such contamination; and (3) methods to monitor the movement of subsurface contaminants.

The projects in this portfolio address a wide range of contaminant types and site characterization problems. Organic contaminants (especially non-aqueous phase liquids) are the subject of 12 projects, compared to three for radionuclides and one for metals; 16 projects focus on site characterization without regard to contaminant type. A majority (23 projects) involve field investigations at contaminated sites. In terms of project objectives, three focus on elucidating contaminant properties, four on elucidating subsurface properties, 13 on the development of invasive characterization techniques, and 12 on the development of noninvasive techniques.

The projects in this portfolio address many of the subsurface problems described in Chapter 2, including aspects of the following topical areas:

- development and testing of noninvasive techniques to identify the distribution of non-aqueous phase liquids in the subsurface;
- development and validation of analytical and modeling tools to be used in subsurface process representation and characterization;
- development of techniques and instruments to determine subsurface parameters that describe flow of water and contaminant transport in the subsurface; and
- noninvasive geophysical techniques and associated analytical techniques to determine subsurface physical parameters.

The portfolio is heavily weighted toward organic contaminants, and there are relatively few projects on metals and radionuclides, which are significant problems at most of the large DOE sites. There are also very few projects that deal with the behavior and transport of contaminants in fractured systems, primarily under unsaturated conditions, or the behavior and transport of contaminants under near-surface conditions (e.g., in near-surface release sites).

Contain

The removal and treatment of contaminants from waste burial grounds is technically difficult, expensive, and could expose workers to radiation and hazardous chemicals. For these reasons, DOE does not plan to fully remediate subsurface contamination at some of its sites. Instead, DOE plans to contain the waste at such sites with surface caps and subsurface barriers to minimize water infiltration and contaminant movement. Remediation of contaminated soil and groundwater at many DOE sites is technically impracticable with current technologies, so DOE plans to monitor this contamination and treat it where necessary, using technologies such as pump-and-treat systems to prevent its further spread.[5] Thus, the availability of robust containment and stabilization technologies will be a key factor in the success of DOE's strategy to manage subsurface contamination.

Given the importance of containment and stabilization technologies to contamination management strategy, the committee would have expected to see a large number of projects on this topic; however, the committee was able to identify only six relevant projects in the portfolio (see Table 3.1). In general, these projects are concerned largely with metals and radionuclides and the kinetics and mechanisms of contaminant retention and release through various processes. Five of the six projects focus on chemical stabilization, one on biological stabilization, and one on physical stabilization.[6] Only one of the six projects has a significant field component.

The committee concluded that there are significant research gaps in the portfolio in this category. These gaps[7] include basic research on the

---

[5]Pump-and-treat systems are used frequently to remediate contaminated groundwater. It involves pumping the contaminated water to the surface for treatment and then reinjecting it. See NRC (1994) for a discussion of this technology.

[6]The current portfolio supports several projects on phytoremediation. These are discussed under the "Remove" category elsewhere in this chapter.

[7]In the context of this analysis, the committee defines a research gap as a deficiency in the number or scope of research projects that address the difficult DOE cleanup problems identified in Chapter 2. The identification of gaps involves a significant element of judgment, especially in interpreting the significance of the subsurface contamination problems now at DOE sites. These cleanup problems and associated knowledge gaps are discussed more fully in Chapter 5.

design, performance, or effectiveness of engineered surface or subsurface barriers, including capillary or resistive barriers, reactive barriers, or hybrid barriers that incorporate biological materials; and research on subsurface processes that address the potential effectiveness of natural barriers in contaminated areas, particularly in the vadose zone.

Remediate

Technologies for in situ treatment and destruction involve the use of engineered or artificially manipulated natural processes to promote the conversion of subsurface contaminants to nonhazardous or less hazardous forms. The committee identified 37 projects in the portfolio (see Table 3.1) that address a wide range of chemical, physical, and biological treatment and destruction processes, including the following:

- bioremediation,[8] including biological interactions, genetic engineering studies, and toxicity studies;
- in situ physical and chemical treatment, including electrochemical processes; filtration; sorption; and reactive subsurface barriers such as metal (Fe, Mn) oxide barriers, including passive or low-maintenance barriers;
- coupled chemical, physical, and biological treatment processes used in parallel or series; and
- elucidation of fundamental subsurface processes that govern the effectiveness of in situ treatment or destruction (e.g., evaluation of the effect of soil heterogeneities on treatment processes).

Projects on organic contaminants comprise the majority of the portfolio (24 of 37 projects), whereas only 10 projects address treatment of radionuclides and 14 address treatment of metals.[9] The committee was able to group the projects into one or more of the following five thematic areas: (1) development of new genetic materials to degrade or alter the chemical composition of DOE's most problematic wastes, including mixed wastes containing radionuclides, heavy metals, and

---

[8]Bioremediation generally refers to the removal of contaminants from soil or water through the metabolic action of living organisms, and the term is commonly used to indicate situations in which humans have interceded to bring about or hasten the biodegradation of contaminant compounds. Although bioremediation can be carried out by any living organisms (e.g., as in phytoremediation), it is usually considered to be a product of the metabolism of microorganisms such as bacteria or fungi.

[9]Some projects address more than one contaminant type.

solvents; (2) elucidation of molecular-level biochemical, geochemical, and biogeochemical processes to degrade or transform selected waste components; (3) taking basic science results to the technology implementation level to develop in situ engineered systems; (4) development of improved analytical methods to allow evaluation of the effectiveness of in situ treatment or destruction; and (5) development of improved understanding of transport processes at all scales in heterogeneous systems that affect the movement of contaminants in the subsurface.

The portfolio defines a fairly coherent research program on in situ treatment and destruction, but there are a number of significant gaps as outlined below, and for some research topics there appears to be duplication of effort. The following observations are, in the committee's view, most significant:

- There is a predominance of projects that address bioremediation relative to projects that address chemical and physical processes.
- Research on treatment and destruction in the vadose zone is underrepresented.
- Research on sensors is bio-oriented and much of it is aimed at tracking the biological "health" of subsurface systems.
- In the bioremediation area, there is an absence of projects covering (1) alternate electron acceptors, including iron and nitrate, and aerobes (the issue of aerobic degradation is important for vadose zone applications); (2) toxicity of some chemical contaminants found at DOE sites to bacteria that could potentially degrade other contaminants; and (3) cellular mechanisms and processes important to the bioremediation of radionuclide and organic contaminants, including the byproducts of microbial degradation activity.
- Understanding what controls the availability of many contaminants to degrading organisms or to reacting chemicals is needed.

Remove

DOE uses the term "hot spot" to refer to significant contaminant source terms in the subsurface that cannot be treated by in situ methods (DOE, 1998d). In lay terms, a hot spot is a distinct high-concentration contaminant anomaly in the subsurface (e.g., a pool of non-aqueous phase liquids trapped in a waste burial ground or a buried 55-gallon drum filled with plutonium-bearing scrap metal). Removal of hot spots involves the physical extraction of the contaminant from the subsurface for ex situ treatment or disposal.

None of the projects in the EM Science Program portfolio have a specific focus on hot spots, however there are seven projects (see Table 3.1) on phytoremediation, an intensely pursued approach to soil cleanup and extractive technology for treatment of hot spots. Research projects include the study of genetic factors controlling the uptake of heavy metals by plants, transport of heavy metals across plant cells, and the ability of plants (poplar trees) to take up and degrade chlorinated hydrocarbons.

Moreover, many of the projects are relevant to improved decision making about whether to contain, stabilize in situ, or extract hot spots for above-ground treatment. For example, some of the projects in the "Other" category discussed later in this chapter are relevant in this regard. Some of the studies in the portfolio on removal and neutralization of contaminants in tank wastes may lead to results useful for treatment of extracted hot spot materials.

Similarly, research projects on locating and quantifying contamination, which were discussed earlier, could make the location and definition of hot spots easier, faster, more accurate, and more economical. Moreover, there are projects in the portfolio that address reactive barriers, bioremediation, in situ vitrification, waste treatment and extraction using electrokinetics, non-aqueous phase liquid migration and pooling, surfactants, adsorption-desorption reactions, and contaminant transport. Many of these projects fall into the "Other" category discussed later in this chapter. The challenge to DOE is to understand and apply the results of this research in dealing with hot spots in reliable and cost-effective ways.

Validate

The Subsurface Contaminants Focus Area defines "Validate" as "validate and verify system performance for regulators and stakeholders" (DOE, 1998d, p. 4). The committee has adopted a somewhat more expansive description that includes confirmation of the effectiveness of remediation processes or strategies. The committee also includes in its definition the validation of conceptual models and the performance of quantitative models of contaminant fate and transport. Under the committee's expanded definition, performance validation is a major factor in regulatory acceptance. It underpins all of DOE's site remediation activities and provides tools and methods to assess the effectiveness of cleanup efforts.

The committee identified nine projects that address the problems in this category (see Table 3.1). Two of these projects address validation of contaminant detection and characterization, three address the validation of fate and transport (i.e., performance of models for fluid flow), and four address remediation effectiveness (i.e., validation of in situ biodegradation or immobilization efforts). The portfolio does not, however, represent a coherent research program in the validate performance area. Notably absent are projects to validate long-term performance of containment systems, including containment barriers. Also missing from the portfolio are projects to develop protocols for validation of conceptual and numerical models of contaminants in the subsurface. The committee believes that validation is a key area for future work by the EM Science Program, as explained in Chapter 5.

Other

The portfolio includes several projects that have indirect but potentially very significant applications to DOE's subsurface contamination problems. In particular, the program is supporting several projects on the biological effects of radiation and hazardous chemicals, including impacts on health and risk (see Table 3.1).[10] Relevant projects fall into the following four thematic areas:

1. effects of radiation and hazardous chemicals on human health and risk (seven projects);
2. effects of contaminants on ecology and ecological risk (three projects);
3. genetic or molecular basis for contaminant effects (four projects); and
4. assessment of monitoring techniques for environmental contaminants (two projects).

None of these projects addresses explicitly the remediation of subsurface contamination, but they are nevertheless relevant to subsurface cleanup efforts because they contribute to the body of science that regulatory agencies use to set cleanup standards and levels.

These projects do not define a coherent research program on biological effects and, in fact, the portfolio of projects could be characterized as meager, given the potential significance of this area on DOE's cleanup efforts.

---

[10]As noted in Chapter 1, the EM Science Program awarded funds for research on low dose radiation in fiscal year 1999.

Discussion And Conclusions

The EM Science Program is by design a "bottoms-up" program in which investigators are encouraged to submit their research ideas to address cleanup problems. In this respect, the program is not unlike other basic research programs operated in DOE's Office of Science and other federal agencies, like the National Science Foundation. Funding decisions are based on the scientific merit of the research proposal and its relevance to DOE problems (see Appendix A). The selection process has resulted in many scientifically meritorious and relevant projects, but there has been a limited opportunity to build coherence. The committee discusses ways to increase coherence in Chapter 5.

The EM Science Program is nevertheless supporting 91 projects focused on subsurface contamination problems[11] and on health and risk effects that are potentially relevant to these problems. It is not unreasonable to expect that the program will attain a critical mass of projects in some problem areas. The purpose of the assessment in this chapter is to determine where these critical masses are present and to identify important gaps in the portfolio that DOE should fill in future competitions. Of course, the committee recognizes that some of the gaps identified may in fact be addressed in other federal research programs and in more recent EM Science Program proposal awards. A discussion of other federal programs is provided in Chapter 4.

The program portfolio in subsurface research has some significant areas of strength. For example, the portfolio has a good selection of projects that address organic contamination problems (50 projects) and that use field-based approaches or a combination, of field-, laboratory-, and modeling-based approaches (38 projects). There appears to be a critical mass of projects in the "Remediation" category, especially for treatment and destruction of organic contaminants through physical, chemical, and biological processes. The committee did observe gaps in the portfolio in this problem area, as noted previously, but these are minor in comparison to gaps in other categories.

The most notable gaps in the portfolio are in the "Contain" and "Validate" categories, two of the most significant problem areas for DOE given its plans to manage much of its subsurface contamination in place. In the "Contain" category the gaps include research on the design, performance, or effectiveness of engineered surface or subsurface barriers. The portfolio in the "Validate" category (9 projects) is lim-

[11]There are 105 projects listed as funded in Table 3.1, but some projects were counted in more than one category. There are 91 separate projects represented by the data in that table.

ited both in terms of depth and breadth of topical coverage. The most notable gaps include research to validate long-term performance of containment systems, including reactive barriers and cover performance, and research to address the validation of conceptual and numerical models of the subsurface and contaminant fate and transport. As noted elsewhere in this report, these are key problems for DOE because they underpin efforts to confirm the effectiveness of and obtain regulatory acceptance for its remediation actions.

There also appears to be a gap in the number of research projects covering radionuclide and metal contamination problems (26 and 31 projects, respectively). As noted in Chapter 2, radionuclide, especially transuranic, contamination is a significant problem, and transuranic contamination is almost exclusively a DOE-owned problem. As will become apparent in the following chapter, these contaminants are receiving relatively little attention in other federal research programs and therefore deserve to be emphasized in future EM Science Program competitions.

4

Research Programs in Other Agencies of Government

As part of its task to formulate recommendations for a long-term research program to address the U.S. Department of Energy's (DOE's) subsurface contamination problems, the committee was asked to consider research already completed or in progress by other federal and state agencies and to identify areas where the Environmental Management (EM) Science Program could make significant contributions (see Sidebar 1.1). The committee partially addressed this task in Chapter 3 by reviewing research that was completed or underway in the program itself. In this chapter, this task is completed by examining research programs in other agencies of government.[1]

The committee gathered information for this review from a variety of sources. The committee received briefings on research programs in five federal agencies at its fourth information-gathering meeting (see Appendix B): Department of Defense (DOD), DOE, Environmental Protection Agency (EPA), National Science Foundation (NSF), and the U.S. Geological Survey (USGS). The purpose of these briefings was to provide an overview of the research programs and to give the committee an opportunity to assess how well these programs were being coordinated. The committee then conducted additional research on these and other programs through electronic searches,[2] followed by contacts with selected program managers.

The committee's initial plan was to summarize the information on other research programs using the organizing scheme shown in Figure

---

[1]Although the statement of task explicitly directs the committee to examine research in "other federal" agencies, the committee has interpreted its mandate to include research in other parts of DOE, especially the Office of Science.

[2]Searches were conducted using the Internet, specifically research databases such as the Federal Information Exchange at http://web.fie.com/fedix/.

3.2, which depicts the Subsurface Contaminants Focus Area's approach to organizing its technology development programs. However, it quickly became clear that such an approach was impractical. In general, the committee found that most other research programs could not neatly be categorized into one or two of the boxes shown in Figure 3.2. In fact, many of the research programs were quite broad in scope, and it was not possible to obtain an accurate picture of the research being sponsored without a detailed review of project portfolios, much like the committee provided in Chapter 3 for the EM Science Program. There simply was not enough time available in this study to do that kind of review for all of the programs discussed in this chapter.

The committee was surprised by the large number of programs that deal either directly or indirectly with subsurface contamination research. Indeed, the committee identified almost 50 programs that could be related at least indirectly to the work of the EM Science Program, not including the programs on health and health effects sponsored by the National Institutes of Health.[3] Thus, to address its task statement, the committee decided to summarize the scope and objectives of these related research programs and to use these descriptions to formulate recommendations for the EM Science Program.

The description of these related research programs is provided in Table 4.1 (located at the end of this chapter), which groups them by agency, and then by program within each agency. The table provides a short description of program scope and objectives;[4] recent funding levels if available;[5] a notation showing whether the program provides intramural or extramural funding;[6] and a web address (if available) where additional information can be obtained. The programs in Table 4.1 are

---

[3]The committee decided to exclude health-related research programs mainly because health research has not been an important component of the EM Science Program. However, the program did focus part of its fiscal year 1999 program competition on low-dose radiation, in cooperation with the DOE-Office of Science's Low Dose Radiation Research Program. This competition was completed while the committee's report was in review. The results from these and other related research programs may have a significant impact on DOE's cleanup program, specifically in establishing the adequacy of DOE's cleanup and containment efforts.

[4]Program information was derived from descriptions provided by the agencies in their program announcements or at their web sites.

[5]Funding amounts are for the entire research programs; only a fraction of the amount listed may be for support of subsurface contamination projects. In most cases it was not possible to separate the subsurface research component.

[6]That is, funding for research conducted within the agency by agency investigators (intramural funding), or funding for research conducted outside of the agency (extramural funding). Extramural funding is typically provided to investigators in academia, national laboratories, industry, or other federal agencies through grants, contracts, and cooperative agreements.

for federal agencies only; the committee was unable to find any significant state-funded basic research programs.[7]

The remainder of this chapter consists of three sections. In the first section, the committee provides a brief review of those research programs that appear to be closely related in terms of scope and objectives to the EM Science Program. The second section provides a short discussion of other programs, and the final section provides some concluding observations.

Closely Related Research Programs

The committee's selection of a research program as closely related to the EM Science Program is based on two somewhat qualitative criteria: (1) the degree to which the program sponsors basic research, as compared to other activities like technology development; and (2) the degree to which the program sponsors research that addresses the topics shown in Figure 3.2. The committee included those programs that it judged had a good match with both criteria.

Of the programs shown in Table 4.1, the committee judges that the following 18 programs in eight federal agencies are closely related in terms of scope and objectives to the EM Science Program:

- *U.S. Department of Agriculture.* **The Environmental Chemistry Laboratory** sponsors intramural and cooperative research on phytoremediation and accelerated microbial degradation of organic compounds and has an annual budget of about $2 million.[8]
- *U.S. Department of Defense.* The Army's **Terrestrial Science Program** sponsors extramural research on experimental, theoretical, and numerical studies on fluid flow and contaminant transport processes in heterogeneous porous media. It has an annual budget of about $1 million.

  The **Naval Research Laboratory** sponsors research on in situ remediation, microbial degradation processes, and environmental monitoring.

  The **Strategic Environmental Research and Development**

---

[7]The committee recognizes that individual states may provide research funding to state agencies and universities for environmental-related basic research, but the committee was unable to identify any state programs that provide state taxpayer dollars at the levels commensurate with the federal agencies listed in Table 4.1.

[8]Unless otherwise noted, the budget numbers given in this chapter are for fiscal year 1999.

Program sponsors extramural research on cleanup, compliance, conservation, and pollution prevention. The program is managed in cooperation with DOE and the EPA and has an annual budget of $59.4 million. About $18.4 million of this budget is directed to cleanup-related research.

- *U.S. Department of Energy.* In DOE's Office of Science, there are several programs in basic energy sciences that sponsor extramural research to understand fundamental physical, chemical, biological, and geological processes (see Table 4.1). Some research sponsored by these programs is relevant to environmental cleanup, but none is focused explicitly on the topical areas shown in Figure 3.2. There appear, however, to be at least two programs in the Office of Science that are directly relevant to the EM Science Program. The Office of Biological and Environmental Research's **Natural and Accelerated Bioremediation Program** sponsors extramural research to understand and apply natural processes to accelerate the biologically enhanced immobilization or degradation of contaminated soil and groundwater.

  In DOE's Office of Environmental Management, the Office of Science and Technology supports a number of applied research, technology development, and technology deployment programs. The overall objective of these programs is to bring new and improved technologies to bear on cleanup of the DOE complex.

  DOE also supports numerous user facilities at several national laboratories (see Table 4.2). Many of these support environmental-related research funded by DOE and other research programs.
- *U.S. Department of Interior.* The U.S. Geological Survey's **Toxic Substances Hydrology Program** funds intramural research on point source contamination in the environment. This program has sponsored 10 field sites around the country (see Sidebar 4.1) to encourage collaborative research among USGS and outside scientists on problems ranging from landfill leachates to mine tailings waste. The use of field sites encourages research collaborations and spreads the costs of expensive monitoring and other observational facilities. The program has an annual budget of about $10 million, and the field sites themselves are made available at no cost to scientists outside the USGS. These scientists must obtain additional funding from their organizations or from other research programs to support the costs of their research projects.
- *U.S. Environmental Protection Agency.* The Office of Research and Development finances a large number of research programs that are directly relevant to the EM Science Program. Almost all these programs are addressing problems of hazardous waste

Table 4.2 U.S. Department of Energy User Facilities

Maintained by Basic Energy Sciences, Division of Materials Sciences
Advanced Light Source, Lawrence Berkeley National Laboratory
Advanced Photon Source, Argonne National Laboratory
Intense Pulsed Neutron Source, Argonne National Laboratory
National Synchrotron Light Source, Brookhaven National Laboratory
Los Alamos Neutron Scattering Center
High Flux Isotope Reactor, Oak Ridge National Laboratory
Stanford Synchrotron Radiation Laboratory, Stanford Linear Accelerator Center
High Flux Beam Reactor, Brookhaven National Laboratory
Materials Preparation Center, Ames Laboratory
Electron Microscopy Center, Argonne National Laboratory
Center for Microanalysis, University of Illinois
National Center for Electron Microscopy, Lawrence Berkeley National Laboratory
Shared Research Equipment Program, Oak Ridge National Laboratory
Surface Modification and Characterization Research Center, Oak Ridge National Laboratory
Combustion Research Facility, Sandia National Laboratory, Livermore, California
Maintained by Basic Energy Sciences, Division of Chemical Sciences
National Synchrotron Light Source, Brookhaven National Laboratory
High Flux Isotope Reactor, Oak Ridge National Laboratory
Radiochemical Engineering Development Center, Oak Ridge National Laboratory
Combustion Research Facility, Sandia National Laboratories, Livermore, California
Stanford Synchrotron Radiation Laboratory, Stanford University
Maintained by the Office of Biological and Environmental Research
The Atmospheric Radiation Measurement Observation Sites (Southern Great Plains, Tropical Western Pacific, and the North Slope of Alaska)
Environmental Molecular Sciences Laboratory, Pacific Northwest National Laboratory
Production Sequencing Facility, Joint Genome Institute, University of California
Mouse Genetics Research Facility, Oak Ridge National Laboratory
Office of Biological and Environmental Research conducts research at the following user facilities
Advanced Light Source, Protein Crystallography Program, Lawrence Berkeley National Laboratory
Advanced Light Source, Soft X-ray Spectroscopy Program, Lawrence Berkeley National Laboratory
Brookhaven High Flux Beam Reactor (neutron crystallography and scattering), Brookhaven National Laboratory
Los Alamos Neutron Science Center (protein crystallography with neutrons), Los Alamos National Laboratory
National Synchrotron Light Source (X-ray crystallography of biological macromolecules and UV spectroscopy), Brookhaven National Laboratory
Oak Ridge High Flux Isotope Reactor (neutron crystallography), Oak Ridge National Laboratory
Stanford Synchrotron Radiation Laboratory (crystallography, spectroscopy, and small-angle scattering of biological molecules), Stanford University
Structural Biology Center (crystallography of biological macromolecules), Argonne National Laboratory

management and cleanup in the nation's civilian sector. The **National Exposure Research Laboratory,** in Research Triangle Park, North Carolina, sponsors research to improve capabilities to locate, characterize, and remediate volatile organic compounds, including dense non-aqueous phase liquids, in subsurface environments. The annual budget is about $3.8 million.

The **National Risk Management Research Laboratory,** in Cincinnati, Ohio, sponsors intramural research on contaminated groundwater and soil and on containment systems. The groundwater research program focuses on source characterization, remediation, and modeling of organic compounds and such metals as arsenic. The annual budget is about $4.2 million. The soil research program covers in situ remediation, including biotreatment, of persistent organic and metal (lead and cadmium) contaminants in soils, sediments, and unsaturated subsurface environments. The annual program budget is about $5.6 million. The

SIDEBAR 4.1 LONG-TERM RESEARCH SITES

The U.S. Geological Survey (USGS) maintains a number of long-term research sites for the study of point source contaminants in the environment. The sites serve as natural laboratories at which scientists conduct experiments and long-term observation. They have proven to be ideal settings for the development of scientific knowledge about the fate and transport of contaminants.

Examples of sites and contaminants studied include tritium from a low-level radioactive waste disposal site in Nevada; sewage effluent from ponds in Massachusetts; oil from a petroleum pipeline break in Minnesota; oxygenated gasoline from buried tanks in South Carolina; creosote effluent from a creosote facility in Florida; mining tailings pond leachate in Arizona; leachate from a landfill in Oklahoma; and organic contaminants from an arsenal in New Jersey. An uncontaminated site in New Hampshire was established to study fracture flow. Some of these sites have been in existence for over a decade. Work at several of the sites was curtailed when scientific interest was satisfied.

Each site is managed by a USGS field scientist who lives and works nearby. This person is responsible for maintaining a stable research site by maintaining good working relations with the land owner, scheduling field research, facilitating the research by helping to provide the technical infrastructure, ensuring that research projects do not interfere with one another, and maintaining the site data base.

A USGS research coordinator is assigned to work with the site manager and to serve as the link between the site and the research community. Knowledgeable about the site environment and the particular contaminant, the coordinator is responsible for making the existence of the site known in the appropriate research communities and to assist the site manager in coordinating science at the site.

The sites have provided fertile environments for scientific study. The prospects of a long-term site with stable scientific management, field assistance, and a long-term database have attracted top scientists in multiple disciplines from academia, government, and the private sector.

containment research program aims to develop new materials and methods for containment of contaminated groundwater and soil; the annual budget is about $1.9 million.

The **National Center for Environmental Research and Quality Assurance** in Washington, D.C., sponsors five hazardous substance research centers in cooperation with universities across the United States. These centers were established under the Comprehensive Environmental Response, Compensation, and Liability Act (the Superfund Act), and their primary funding comes from the EPA (about $8.9 million in fiscal year 1999), with additional funding from other federal agencies, universities, state agencies, and the private sector. These centers have research foci that are related directly to the EM Science Program:

1. The Great Lakes/Mid-Atlantic Center sponsors research on remediation of hazardous organic compounds found in soils and groundwater. The University of Michigan leads the three-institution consortium.
2. The Great Plains/Rocky Mountain Center sponsors research on soils and mining wastes contaminated with organic chemicals and heavy metals. Kansas State University leads a fourteen-institution consortium.
3. The South/Southwest Center sponsors research on in situ detection, mobilization, and remediation of contaminated sediments. Louisiana State University leads the three-institution consortium.
4. The Western Center sponsors research on groundwater cleanup and site remediation for organic solvents, hydrocarbons and derivatives, and heavy metals. The center is a cooperative activity involving Oregon State University and Stanford University.
5. The Gulf Coast Center sponsors research on hazardous substance response and waste management. The center is a cooperative activity involving eight universities.

The National Center for Environmental Research and Quality Assurance, in collaboration with DOE, the Office of Naval Research, and NSF, also sponsors a program in **bioremediation** that seeks to understand the chemical, physical, and biological processes that influence the bioavailability and release of chemicals in soil, sediments, and groundwater. The annual funding for this program is about $1 million.

- *U.S. Department of Health and Human Services.* The National Institute of Environmental Health Sciences sponsors a joint program with the EPA on the **Superfund Hazardous Substances Basic Research Program,** which has an annual budget of about $37 million. This program supports research to understand hazardous waste exposure risks and to support the development of site remediation technologies.
- *National Science Foundation.* Like DOE's Office of Science, the NSF sponsors several extramural research programs to understand fundamental physical, chemical, biological, and geological processes (see Table 4.1). Some of these programs sponsor research that is directly relevant to environmental cleanup, but none is focused explicitly on the topical areas shown in Figure 3.2. There are at least two programs in the NSF that appear to be directly relevant to the EM Science Program. The **Civil and Mechanical Systems Program** sponsors basic engineering research, including geotechnical research on materials, containment systems, remediation, and modeling. The annual funding for this program is about $59.5 million.[9] The NSF also sponsors a cross-directorate program on **Environmental Geochemistry and Biogeochemistry,** whose goal is to improve fundamental knowledge of chemical processes that control the behavior and distribution of inorganic and organic materials in the environment. The annual funding for this program is about $4.8 million.
- *U.S. Nuclear Regulatory Commission.* This agency is charged with regulating the production, use, and disposal of radioactive byproduct materials; it sponsors research through the **Center for Nuclear Waste Regulatory Analysis** in San Antonio, Texas. The Geohydrology and Geochemistry Section in this center sponsors research on surface and subsurface hydrology related to the transport and fate of contaminants.

Other Research Programs

Table 4.1 lists a number of other programs that sponsor research that is less directly relevant to the EM Science Program; nevertheless, these programs support research that may in the long term support the DOE cleanup effort. The basic research programs in DOE's Office of Science and the National Science Foundation, which were mentioned in the last section, are good examples. They sponsor research that will increase

---

[9]Only a portion of this total is for geotechnology-related research.

the basic knowledge pool that can be accessed by the EM Science Program and its researchers. Many of the researchers who receive EM Science Program funding are also being or have been supported by one or more basic research programs in DOE and NSF.

There is another group of programs in Table 4.1 that has some relevance for the EM Science Program and DOE's overall cleanup efforts, namely, the programs that support risk analysis and risk assessment research. Risk assessment is an important step in the remediation process, as will be shown in Chapter 5, and the EM Science Program is now supporting several projects that address risk-related topics (see Chapter 3). There are several research programs in Table 4.1 that address various aspects of hazard and risk assessment:

- The EPA's National Center for Environmental Assessment sponsors two research programs in this area, one on **Superfund health risk assessment,** with an annual budget of $2.1 million, and a second on **Superfund ecological risk assessment,** which has an annual budget of $1.0 million.
- The EPA's National Center for Environmental Research and Quality Assurance, in cooperation with the National Institute of Environmental Health Sciences, sponsors a program on **complex mixtures** that focuses on the mechanistic basis for chemical interactions on biological systems. The annual budget is about $2.7 million.
- As mentioned in the last section, the National Institute of Environmental Health Sciences sponsors a joint program with EPA on the **Superfund hazardous basic research.** One of the objectives of this program is to understand hazardous waste exposure risks.

Discussion

In responding to its statement of task, the committee attempted to survey research completed or underway in other federal or state agencies that it could use in formulating a long-term research agenda for the EM Science Program. The committee attempted to identify those research programs that seemed to be most closely related to the EM Science Program and to gain a general understanding of research objectives. The committee believes that this review has provided enough information to make the following five observations that will be used to formulate recommendations for the long-term research agenda presented in Chapter 6.

1. The federal government is a major sponsor of basic research related either directly or indirectly to environmental problems. The committee identified almost 50 research programs in its survey (see Table 4.1). If health-related research programs were included in the committee's survey, the number would be much higher.
2. There are a large number and variety of programs across the federal government that support research of direct relevance to the EM Science Program and DOE's cleanup problems. The committee identified 18 such research programs.
3. There appears to be significant overlap in scope among some of the programs identified in this analysis, judging from the program descriptions given in Table 4.1. Overlap is not necessarily undesirable, but it is not clear whether there is an effective mechanism to coordinate these programs. There are some notable exceptions to this generalization, especially for those programs listed in Table 4.1 that are jointly managed by several agencies (e.g., the Strategic Environmental Research and Development Program, which is managed by the DOD in cooperation with DOE and the EPA.).
4. Many of the 18 directly relevant programs identified in point 2 above focus on hazardous chemicals, and to a lesser extent on heavy metals. There appear to be few programs that address radionuclide contamination outside DOE.
5. Many of the 18 directly relevant programs also focus on remediation, and especially bioremediation. Other remediation approaches and other important research topics related to environmental cleanup (e.g., contaminant location and characterization in the subsurface) appear to be receiving less attention.

The committee believes there would be value added to the federal government's basic research on environmental problems if there was better coordination among its research programs, especially the mission-directed programs. The committee sees an opportunity for EM Science Program managers to promote and foster such coordination.

There are many good coordinating mechanisms that have been used elsewhere in the federal government that could be adapted to coordinate these mission-directed environmental research programs. These range from formal coordinating mechanisms like the Federal Remediation Technologies Roundtable[10] to more informal mechanisms like

[10]The Federal Remediation Technologies Roundtable is an interagency coordinating group comprising representatives of federal agencies with hazardous waste cleanup responsibilities. The roundtable provides a forum for information

periodic meetings of interested program managers, or even joint sponsorship of field research sites to address specific contamination problems. Regardless of the mechanisms, however, the objective should be to improve communication among federal program managers, reduce unnecessary duplication and overlap among programs, and help program managers focus their resources on those problems that provide the greatest challenges to the nation's environmental cleanup efforts.

---

exchange and joint action concerning the development and demonstration of innovative technologies for hazardous waste remediation. Additional information is available at http://www.frtr.gov.

TABLE 4.1 Relevant Research in Federal Agencies

(Note: Funding information is for fiscal year 1999 unless otherwise noted. The funding levels are for comparison purposes only and are approximate figures based on the best available data. In some cases reliable data could not be acquired. This list is not comprehensive.)

Section, Department, or Division	Program	Sub-program	Funding Level	Intra- or Extramural	Web Site for Source and More Information
U.S. DEPARTMENT OF AGRICULTURE					
Agricultural Research Service	National Program on Natural Resources and Sustainable Agriculture	Water Quality and Water Management	All soil and water within ARS: FY 99: $86M FY 98: $86M	Intra	http://www.nps.ars.usda.gov/programs/201b.htm
The primary mission of this national program is twofold: to develop innovative concepts for determining the movement of water and its associated constituents in agricultural landscapes and watersheds, and to develop new and improved practices, technologies, and strategies for managing the nation's agricultural water resources. These advances will provide food and fiber producers, local communities, and resource management agencies with tools to improve water conservation and water use efficiency in irrigated and dryland agriculture, enhance water quality, protect rural communities from the ravages of floods and droughts, and prevent the degradation of landscapes, aquatic ecosystems, and stream corridors.					
Agricultural Research Service	Soil, Water, and Air Quality/ Environmental Chemistry Laboratory	Cooperative Research: metals in soil; metals/plant uptake	$2M	Both	http://hydrolab.arsusda.gov/ecl/CRIS%20Reports/metalsin1.htm
This program targets toxic trace elements, organic agrochemicals, and radionuclides. The main approaches include: a) use of specially designed biosolid compost to reduce the bioavailability of toxic trace elements, b) phytoremediation using hyperaccumulator plants or plants that convert toxic elements to nontoxic forms, and c) accelerated microbial degradation of organic compounds. Funding for this program is approximately $1.5M of agency funding and approximately $0.5M in grant funding per year.					
U.S. DEPARTMENT OF COMMERCE					
National Institute of Standards and Technology	Chemical Science and Technology Laboratory	Divisions (approximate funding levels based on FY98 projections): Biotechnology (~$10.3M)	FY 98: Approximately $58M	Intra	http://www.nist.gov/public_affairs/guide/cstpage.htm

		Process Measurements (~$10.6M) Surface and Microanalysis Science (~$8.4M) Physical and Chemical Properties (~$13.3M) Analytical Chemistry (~$13.9M)			

The Chemical Science and Technology Laboratory performs cutting-edge research in measurement science; develops and maintains measurement methods, standards, and reference data; and develops models for chemical, biochemical, and physical properties and processes. The lab provides these capabilities to enhance U.S. industry's productivity and competitiveness; ensure equity in trade; and improve public health, safety, and environmental quality.

National Institute of Standards and Technology	Materials Science and Engineering Laboratory	Divisions (approximate funding levels based on FY98 projections): Materials (~$5.7M) Metallurgy (~$10.9M) Polymers (~$8.9M) Ceramics (~$11.9M) NIST Center for Neutron Research (~$14.9M)	FY 98: Approximately $50M	Both	http://www.msel.nist.gov/

The Materials Science and Engineering Laboratory provides technical leadership and participates in developing the measurement and standards infrastructure related to materials critical to U.S. industry, academia, government, and the public. Materials science and engineering programs cover a full range of materials issues from design to processing to performance. A unifying aim is to acquire the knowledge and tools needed for intelligent manufacturing methods with real-time automated process controls. Separate research initiatives address ceramics, metals, polymers, composites, and superconductors. This research supports efforts of U.S. industry to develop reliable, low-cost manufacturing methods for tailor-made materials and products with superior properties.

TABLE 4.1 Continued

(Note: Funding information is for fiscal year 1999 unless otherwise noted. The funding levels are for comparison purposes only and are approximate figures based on the best available data. In some cases reliable data could not be acquired. This list is not comprehensive.)

Section, Department, or Division	Program	Sub-program	Funding Level	Intra- or Extramural	Web Site for Source and More Information
National Institute of Standards and Technology	Physics Laboratory	Divisions (funding levels are approximate based on projections for FY98: Electron & Optical Physics (~$5.8M) Atomic Physics (~$8.1M) Optical Technology (~$11.1M) Ionizing Radiation (~$6.8M) Time and Frequency (~$8.4M) Quantum Physics (~$5.8)	FY98: Approximately $48.3M	Both	http://physics.nist.gov/

The Physics Laboratory is one of the major operating units of the National Institute of Standards and Technology. Its mission is to support U.S. industry by providing measurement services and research for electronic, optical, and radiation technologies. The laboratory pursues directed research in the physical sciences; develops new physical standards, measurement methods, and data; conducts an aggressive dissemination program; and collaborates with industry to commercialize inventions and discoveries. Programs span the full range from tests of fundamental postulates of physics through generic technology to the more immediate needs of industry and commerce. Its constituency is broadly distributed throughout academia, government, and industry.

U.S. DEPARTMENT OF DEFENSE

Air Force	Air Force Materiel Command/Air Force Office of Scientific Research/Air Force Research Laboratory	Chemistry and Life Sciences: Bio-environmental Science	Approximately $5M annually ($2M intra/ $3M extra)	Both	http://web.fie.com/ htdoc/fed/afr/afo/any/text/ any/rib99-1.htm#6

The Bioenvironmental Science Program supports toxicology-related research that investigates the interactions of biological systems with non-ionizing radiation and chemicals of interest to the Air Force. Air Force operations use physical and chemical agents such as non-ionizing radiant energies (radio frequency radiation, microwaves, and laser light), heavy metals (chromium and cadmium), and various chemicals that constitute fuels, propellants, and lubricants of interest to the Air Force that can be potentially harmful to military and civilian personnel, to the surrounding populace, or to the environment. Exposure to these agents may be a direct result of their use during Air Force operations and maintenance and, in the case of chemicals, may also occur indirectly as a result of leaky storage containers that contaminate waste streams, ground water, and soil. The Air Force supports basic research to understand the biological effects of these agents, their mechanisms of toxicity, and the use of experimental and computational modeling in assessing their potential health risks.

Air Force	Air Force Materiel Command/Air Force Office of Scientific Research/Air Force Research Laboratory	Materials and Manufacturing Directorate/ Biotechnology	Approximately $300,000 annually	Intra	http://www.ml.afrl.af.mil/ divisions/mlq/wud_bt.html

The Biotechnology Division conducts in-house research in biodegradation, specifically strategies to predict, enhance, or prevent microbial attack on materials and in biocatalysis to discover and enhance novel metabolic processes for use in synthesis of materials. This program focuses on mitigating the Air Force's industrial hazardous waste streams, and reducing or preventing environmental contamination at Air Force and other Department of Defense sites. This will minimize the high costs and toxic byproducts of conventional chemistry for materials production.

Army	Army Materiel Command/Army Research Laboratory/ Army Research Office	Mechanical and Environmental Sciences Division: Terrestrial Sciences Program	$4.6M[a]	Extra	http://www.aro.army.mil/ research/baa99-1/baa99.htm#es

The Terrestrial Sciences Program is concerned with the impact of Earth's surficial environment on Army activities. Program interests cover a broad spectrum, ranging from terrain characterization and analysis, to military engineering and mobility considerations under combat conditions, to environmental management and stewardship. Primary emphasis is directed toward understanding the behavior of the land surface and the near-surface environments, understanding the natural processes operating on and in these domains, and modeling these environments for predictive and simulation purposes. Special emphasis is given to the need to better understand, model, and predict those conditions that are most extreme, dynamic, or restrictive to systems performance or military operations. The three areas of current interest to the program are terrain properties and characterization, terrestrial processes and landscape dynamics, and terrestrial system modeling and model integration. In all cases, the emphasis is on basic research.

TABLE 4.1 Continued

(Note: Funding information is for fiscal year 1999 unless otherwise noted. The funding levels are for comparison purposes only and are approximate figures based on the best available data. In some cases reliable data could not be acquired. This list is not comprehensive.)

Section, Department, or Division	Program	Sub-program	Funding Level	Intra- or Extramural	Web Site for Source and More Information
Army Materiel Command/ Army Research Laboratory/Army Research Office	Mechanical and Environmental Sciences Division: Terrestrial Sciences Program	Special Programs: Center for Multiphase Fluid Flow and Contaminant Transport	Approximately $1M annually	Extra[b]	http://www.aro.army.mil/ arorev/enviroma.htm http://cmr.sph.unc.edu/CMR/ http://cmr.sph.unc.edu/CMR/
The research team is engaged in a variety of experimental, theoretical, and numerical studies directed toward understanding the fundamentals of fluid flow and contaminant transport processes in heterogeneous, multiphase porous media systems. The objectives are: (1) to observe fluid flow and contaminant transport through controlled laboratory experiments at a variety of scales; (2) to develop and apply numerical models to simulate flow and transport; and (3) to theoretically evaluate the practical consequences and effects of flow and transport phenomena in the context of the dual goals of improved modeling capabilities and practical procedures for and efficiency in environmental restoration. The center is at the University of North Carolina and supported by multiple organizations (Cray Research, Inc.; German Academic Exchange Service; Superfund Basic Research Program; North Carolina Supercomputing Center; U.S. Army Research Office; U.S. Army Waterways Experiment Station; and U.S. Department of Energy). Research undertaken by the University of North Carolina, Chapel Hill team has been coordinated and shared with Army researchers at the U.S. Army Corps of Engineers Waterways Experiment Station. These efforts have resulted in approaches and computer codes that are readily applicable to field applications, and have been made available to the Department of Defense Ground water Model System Program.					
Army	U.S. Army Corps of Engineers/Waterways Experiment Station Vicksburg, Mississippi	Environmental Laboratory/Fate and Effects Research	Approximately $4M annually	Intra[c]	http://www.wes.army.mil/ el/envrest.html
Fate and effects research began in the early 1970s to support the Corps dredging programs, which included development of testing manuals for sediment contaminants. This was quickly expanded to waste characterization, environmental monitoring, numerical modeling, and physical and biological processes. Results of this research were used to assist with waste disposal practices and ground water monitoring at such facilities as Aberdeen Proving Grounds, Rocky Mountain Arsenal, and Pine Bluff Arsenal. Environmental fate and effects research at the station encompasses a variety of programs in support of Department of Defense agencies, other federal agencies, and various state agencies. This environmental R&D was developed from Corps civil works programs with major reimbursable work from military and civilian sponsors. R&D has frequently been implemented in a cooperative environment with other federal agencies (such as the U.S. EPA) to provide uniform guidance to sponsors.					

<table>
<tr><td>Army</td><td>U.S. Army Corps of Engineers /Waterways Experiment Station</td><td>Environmental Laboratory/Dredging Operations and Environmental Research/Contaminant Sediments Focus</td><td>Not Available</td><td>Intra</td><td>http://www.wes.army.mil/el/dots/doer/cs.html</td></tr>
<tr><td colspan="6">This program examines the most commonly considered alternatives for contaminated sediments, which are placement in confined disposal facilities (CDFs) and capping, an option for containment in subaqueous sites. CDFs are located on land or in areas of relatively sheltered water. Many CDFs are near closure, and future locations may include non-traditional areas such as offshore. Treatment to reclaim CDF capacity may be promising for certain sites. Capping has significant potential as a disposal alternative, but issues related to its long-term effectiveness and potential application to deeper waters or high-energy environments require additional environmental investigation. This program will address high priority research needs aimed at reducing costs associated with screening and assessing potential impacts of contaminants and increasing the reliability and acceptability of CDF and capping options for management of contaminated sediments.</td></tr>
<tr><td>Defense Advanced Research Projects Agency (DARPA)</td><td>Defense Sciences Office (Joint program with the Office of Naval Research)</td><td>Chemical Plume Tracing</td><td>Approximately $15M</td><td>Extra</td><td>http://www.darpa.mil/DSO/solicitations/RA98-09/S/additional.html

http://www.darpa.mil/DSO/solicitations/RA98-09/S/cbd.html</td></tr>
<tr><td colspan="6">Many animals (e.g., lobsters, moths, and dogs) are capable of tracking chemical plumes to their source. It is believed that a careful understanding of how animals accomplish this task will lead to the development of artificial plume tracing systems that can be employed to locate the source of chemical plumes for a variety of applications, including the location of land mines (using plumes in air) and underwater unexploded ordnance (using plumes in water).</td></tr>
<tr><td>Defense Advanced Research Projects Agency (DARPA)</td><td>Defense Sciences Office</td><td>Controlled Biological Systems</td><td>Not available</td><td>Extra</td><td>http://www.darpa.mil/DSO/solicitations/BAA98-07/S/cbd.htm</td></tr>
<tr><td colspan="6">The principal objective of this program is to control, influence, and monitor distributed biological systems. Applications of interest include controlling the distribution of biological systems for real-time monitoring of individuals or populations of organisms (e.g., swarms, hives, dens, schools) to seek out and collect information in the environment (air, land, or water) about agents of harm, including chemical or biological weapons and unexploded ordnance. The program will explore the control of biological systems as first warning systems for predicting human health risk. Application of controlled biological systems could include mapping agent concentration and distribution in potentially contaminated air, land, and water, and countermeasure delivery or intelligence information gathering in hostile or inaccessible environments. All program aspects are for defensive purposes only. Other applications could involve controlling the distribution of pest organisms to improve operational environments for troops. To accomplish this objective the program will seek to monitor and use the sensory signals (e.g., chemical, visual, thermal, acoustic, other) employed by biological organisms to forage and reproduce in their environment. Researchers also seek to develop revolutionary methods to interface with individuals or populations of biological systems as they distribute in the environment. The Controlled Biological Systems Program seeks major technology innovation in new concepts for "plugging into" the signals used by biological organisms and using them to direct distribution of biological systems and to collect environmental information.</td></tr>
</table>

TABLE 4.1 Continued

(Note: Funding information is for fiscal year 1999 unless otherwise noted. The funding levels are for comparison purposes only and are approximate figures based on the best available data. In some cases reliable data could not be acquired. This list is not comprehensive.)

Section, Department, or Division	Program	Sub-program	Funding Level	Intra- or Extramural	Web Site for Source and More Information
Navy	Office of Naval Research/ Naval Research Laboratory	Materials Science and Component Technology Directorate/ Center for Bio-Molecular Science and Engineering	Not available[d]	Both	http://cbmsews1.nrl.navy.mil/ http://heron.nrl.navy.mil/contracts/99baa/951.htm http://heron.nrl.navy.mil/contracts/99baa/924.htm
The Center for Bio-Molecular Science and Engineering conducts multidisciplinary research in biotechnology, using the techniques of modern molecular biology, biophysics, chemistry, microelectronics, and engineering to fabricate biosensors, biomaterials, and advanced systems. Current research areas include (1) biophysical chemistry of membranes; (2) research into biosensors, including construction of novel devices, accessories for automated reagent delivery, production of biomolecular recognition elements, or configuration of bioassays for integration into the sensor (targets of detection include explosives, pollutants, pathogens, toxic agents, and hazardous chemicals); (3) genetic engineering of biomaterials; (4) self-assembled films and patterns for advanced material development; (5) liposomal delivery of vaccines and pharmaceuticals; and (6) physical characterization of thin films and surfaces.					
Navy	Office of Naval Research/ Naval Research Laboratory	Chemistry Division Environment and Biotechnology Branch	Approximately $100M annually	Intra	http://chemdiv-www.nrl.navy.mil/index.html
This Environment and Biotechnology Branch has broad management responsibility for biotechnology programs within the NRL and the DOD. The program manager serves as expert consultant and advisor in biotechnology to the Center for Biomolecular Science and Engineering, the division superintendent, to the associate director of research for materials science and component technology, and to the director of defense research and engineering in the Office of the Under Secretary of Defense. Additionally, the program manager serves as a Navy/DOD representative or liaison to numerous interagency committees. The program is divided into medical and nonmedical (or materials) biotechnology. This technology has the potential to produce new vaccines and therapies, new structural materials, and practical solutions to waste site remediation. Thrust areas are biosensors, bioprocessing, materials, biomicroelectronics, and bioremediation. This program is a DOD critical technology and is considered a growth area.					

Navy	Office of Naval Research/ Naval Research Laboratory	Chemistry Division Chemical Dynamics and Diagnostics Branch Environmental Quality Science Section	Not available	Intra	http://pony.nrl.navy.mil/eqshome.html http://chemdiv-www.nrl.navy.mil/index.html

The Environmental Quality Sciences Section conducts basic and applied research in (1) the development and implementation of in situ remediation treatment strategies for contaminated soil, sediment, and groundwater, using a combination of laboratory, mesocosm, and field-scale studies; (2) the use of microbial processes, such as biodegradation, bioemulsification, and mineral cycling to reduce contaminants in waste streams and environmental contamination of site materials; (3) the characterization and monitoring of ecological parameters involving nutrient cycling, food web dynamics, ecosystem characterization, and resource management to enhance pollution control and optimize environmental recovery and restoration; (4) biological-based sensors for chemical, biological, and environmental quality monitoring; (5) the use of mass spectrometry and related techniques to develop instrumentation for real-time environmental analysis, and to detect, identify, and structurally characterize biomolecules, metal-biomolecule complexes, and other materials; and (6) study the fundamental processes desorption, ionization, fragmentation, and gas-phase reactions of large ions.

Office of the Secretary of Defense	Strategic Environmental Research and Development Program	Strategic Environmental Research and Development Program	FY 99: $59.4M (31% for cleanup research)	Extra	http://www.serdp.gov http://www.serdp.gov/funding/Funding_Process.htm

The Strategic Environmental Research and Development Program is the Department of Defense's corporate environmental research and development program, planned and executed in full partnership with the Department of Energy and the Environmental Protection Agency, with participation by numerous other federal and non-federal organizations. Within its broad areas of interest, the program focuses on cleanup, compliance, conservation, and pollution prevention technologies.

U.S. DEPARTMENT OF ENERGY

Office of Environmental Management & Office of Science	Environmental Management Science Program		$47M	Extra	http://www.em.doe.gov/science/

The Environmental Management Science Program is a collaborative program of the DOE's Office of Environmental Management and Office of Science (formerly the Office of Energy Research) that identifies long-term basic science research needs and targets the research and development to EM's problems as a key to developing innovative and cost-effective cleanup methods. The Subsurface Contaminants Focus Area relies on the Office of Science and the Environmental Management Science Program to answer basic science questions regarding contaminant transport, soil interaction, and sorption to underground substrates.

TABLE 4.1 Continued

(Note: Funding information is for fiscal year 1999 unless otherwise noted. The funding levels are for comparison purposes only and are approximate figures based on the best available data. In some cases reliable data could not be acquired. This list is not comprehensive.)

Section, Department, or Division	Program	Sub-program	Funding Level	Intra- or Extramural	Web Site for Source and More Information
Office of Environmental Management	Office of Science and Technology	Technology development and deployment	FY 99: $187.2M	Extra	http://ost.em.doe.gov/ifd/ost/programs.htm http://www.cfo.doe.gov/budget/00budget/highlite/highlite.pdf
The Office of Science and Technology supports a full range of technology research, development, and deployment activities focused on DOE cleanup. The programs are structured around problem "focus areas" (deactivation and decommissioning; mixed wastes; subsurface contaminants; and tanks) and crosscutting issues (characterization, monitoring and sensor technology; efficient separations; industry programs; robotics; and integrated process analysis).					
Office of Science	Basic Energy Sciences	Chemical Sciences	FY 98: $196M	Extra	http://www.er.doe.gov/production/bes/Division.htm#chemical
The Division of Chemical Sciences supports basic research at universities, DOE national laboratories, and other appropriate organizations for the purpose of providing the knowledge required to develop energy technologies to meet national goals of energy efficiency, public health and safety, environmental protection and restoration, and conservation of natural resources. Projects supported are selected for relevance to these goals and for scientific excellence as judged by peer review.					
Office of Science	Basic Energy Sciences	Engineering & Geosciences	FY 98: $40M FY 99: $42M[e]	Both	http://www.er.doe.gov/production/bes/Division.htm#EngGeo
The Division of Engineering and Geosciences conducts research in two disciplinary areas: engineering and geosciences. In engineering research, the goals are to extend the body of knowledge underlying current engineering practice to create new options for improving energy efficiency and to broaden the technical and conceptual knowledge base for solving the engineering problems of energy technologies. In geosciences research, the emphasis is on fundamental knowledge of the processes that transport, concentrate, emplace, and modify the energy and mineral resources and the byproducts of energy production. The research supports existing energy technologies and strengthens the foundation for the development of future energy technologies. Ultimately the research impacts control of industrial processes to improve efficiency and reduce pollution, to increase energy supplies, and to lower cost and increase the effectiveness of environmental remediation of polluted sites.					

Office of Science	Basic Energy Sciences	Energy Biosciences	FY 98: $26.7M FY 99: $29.8M	Both	http://www.er.doe.gov/production/bes/Division.htm#biosciences

The Division of Energy Biosciences supports research to provide a basic understanding of the biological phenomena associated with the capture, transformation, storage, and use of energy. Research on plants and nonmedical microorganisms focuses on a range of biological processes, including photosynthesis, bioenergetics, primary and secondary metabolism, the synthesis and degradation of biopolymers such as lignin and cellulose, anaerobic fermentations, genetic regulation of growth and development, thermophily (e.g., bacterial growth under high temperature), and other phenomena with the potential to impact biological energy production and conversion. The research is fundamental and is selected to broadly support DOE goals and objectives in energy production, environmental management, and energy conservation.

Office of Science	Basic Energy Sciences	Materials Sciences	FY 98: $381M FY 99: $414M[f]	Both	http://www.er.doe.gov/production/bes/Division.htm#materials

The Division of Materials Sciences supports basic research in condensed matter physics, metals and ceramics sciences, and materials chemistry. This research seeks to understand the atomistic basis of materials properties and behavior and how to make materials perform better at acceptable cost through new methods of synthesis and processing. Research is supported in corrosion, metals, ceramics, alloys, semiconductors, superconductors, polymers, metallic glasses, ceramic matrix composites, non-destructive evaluation, magnetic materials, surface science, neutron and X-ray scattering, chemical and physical properties, and new instrumentation. Ultimately the research leads to the development of materials that improve the efficiency, economy, environmental acceptability, and safety in energy generation, conversion, transmission, and use. These material studies affect developments in numerous areas such as solar energy conversion, transportation, electric power production, and petroleum refining.

Office of Science	Basic Energy Sciences	National User Facilities	Both	http://www.er.doe.gov/production/bes/BESfacilities.htm http://www.er.doe.gov/production/ober/facilities.html

The Office of Basic Energy Sciences plans, constructs, and operates 18 major scientific user facilities to serve researchers at universities, national laboratories, and industry. These facilities enable the acquisition of new knowledge that often cannot otherwise be obtained. Each year, thousands of scientists conduct experiments at the user facilities. Other researchers collaborate with these users and analyze the data from the experiments to publish new scientific findings in peer-reviewed journals. See Table 4.2 for a list of these facilities.

TABLE 4.1 Continued

(Note: Funding information is for fiscal year 1999 unless otherwise noted. The funding levels are for comparison purposes only and are approximate figures based on the best available data. In some cases reliable data could not be acquired. This list is not comprehensive.)

Section, Department, or Division	Program	Sub-program	Funding Level	Intra- or Extramural	Web Site for Source and More Information
Office of Science	Biological and Environmental Research	Environmental Sciences/ Environmental Remediation Research/ NABIR & EMSL[g]	FY 98: $54M FY 99: $56M	Both	http://www.er.doe.gov/production/ober/ESD_top.html http://www.er.doe.gov/production/ober/EPR/nabir.html http://www.lbl.gov/NABIR/ http://www.emsl.pnl.gov:2080/ http://www.er.doe.gov/production/ober/facilities.html

The environmental remediation research portfolio is focused on developing an understanding of the fundamental physical, chemical, geological, and biological processes that must be marshaled for the development and advancement of new, effective, and efficient processes for the remediation and restoration of the nation's nuclear weapons production sites. A primary effort is the Natural and Accelerated Bioremediation Research (NABIR) program, which is a basic research program that seeks to provide the scientific understanding needed to make use of natural processes and to develop methods to accelerate these processes for bioremediation of contaminated subsurface sediments and ground water at DOE facilities. Funding for the operation of the Environmental Molecular Sciences Laboratory (EMSL), the only national scientific user facility focused on DOE's environmental missions, is a key part of the commitment to environmental remediation.

Section, Department, or Division	Program	Sub-program	Funding Level	Intra- or Extramural	Web Site for Source and More Information
Office of Science	Office of Biological and Environmental Research	Life Sciences/ Microbial Genome Program		Both	http://www.er.doe.gov/production/ober/microbial.html

The focus of the Microbial Genome Program is to develop the ability to sequence the genetic material of microbial organisms. This will provide detailed genetic information on microorganisms with importance to the environment, energy production, and other important applications. The program, spun-off from the Human Genome Program in 1994, is already providing complete sequence information on key microorganisms. This effort will enable the scientific community to make unique side-by-side comparisons of complete genetic information from microorganisms with similar attributes.

U.S. DEPARTMENT OF THE INTERIOR

U.S. Geological Survey | Water Resources | National Research Program (several related current research topics) | | Intra | http://water.usgs.gov/nrp/index-areas.html

The National Research Program conducts basic and problem-oriented hydrologic research in support of the mission of the U.S. Geological Survey. Relevant hydrologic information is available today to assist the nation in solving its water problems because of a conscious decision made in years past to invest in research. The program has been designed to encourage pursuit of a diverse agenda of research topics aimed at providing new knowledge and insights into varied and complex hydrologic processes that are not well understood. The research emphasis changes through time, reflecting the emergence of promising new areas of inquiry and the demand for new tools and techniques with which to address water resources issues.

U.S. Geological Survey | Water Resources | Toxic Substances Hydrology (Toxics) Program | $10M | Intra | http://toxics.usgs.gov/toxics/

This program's goal is to provide scientific information on toxic substances in subsurface and ground waters for use in avoiding human exposure, developing effective remedial strategies, and preventing further contamination. Funded research categories are point-source groundwater contamination; nonpoint source contamination; and hardrock mining contamination

U.S. Geological Survey | Biological Resources Division | National Programs: Biomonitoring of Environmental Status and Trends (BEST) | $2.1M | Intra | http://www.best.usgs.gov/

The Biomonitoring of Environmental Status and Trends Program has been designed to identify and understand the effects of environmental contaminants on biological resources, particularly those resources under the stewardship of the Department of the Interior. The program provides sound scientific information to be used proactively to prevent or limit contaminant-related effects on biological resources. Primary goals of the program are to (1) determine the status and trends of environmental contaminants and their effects on biological resources; (2) identify, assess, and predict the effects of contaminants on ecosystems and biological populations; and (3) provide summary information in a timely manner to managers and the public for guiding conservation efforts. The program uses different approaches to goal achievement, including a combination of field biomonitoring methods and information assessment tools to examine contaminant issues at national, regional, and local levels. Biological research is essential for support of these program components, including development of biomonitoring tools and establishment of links between biological responses and exposure to contaminants. Research conducted at the Environmental and Contaminants Research Center and Patuxent Wildlife Research Center has contributed essential tools to the program. In addition, data management, geographical information systems, and Internet capabilities developed at the Midcontinent Ecological Science Center have been critical to program implementation.

TABLE 4.1 Continued

(Note: Funding information is for fiscal year 1999 unless otherwise noted. The funding levels are for comparison purposes only and are approximate figures based on the best available data. In some cases reliable data could not be acquired. This list is not comprehensive.)

Section, Department, or Division	Program	Sub-program	Funding Level	Intra- or Extramural	Web Site for Source and More Information
U.S. ENVIRONMENTAL PROTECTION AGENCY					
Office of Research and Development	National Exposure Research Laboratory (Research Triangle Park, N.C.)	Surface/ Subsurface Characterization and Sampling	\$3.8M	Intra	http://www.epa.gov/crdlvweb/
Inadequate site characterizations and a lack of knowledge of surface and subsurface contaminant distributions (particularly volatile organic compounds and NAPLs) hinder our ability to make good decisions on remediation options and conduct adequate cleanup efforts. Noninvasive geophysical techniques can provide methods for subsurface site characterization. Future effort is being directed toward experiments at the unique field test facility for evaluating these geophysical technologies under controlled DNAPL spill conditions. This lab will also be used to evaluate other ground water sampling methods and designs. Research is being conducted to develop and implement state-of-the-science approaches to improved soil sampling techniques, measurement design and geostatistics, and data analysis through chemometric and robust statistical techniques. Improvements in all aspects of soil sampling are being investigated to quantify and reduce or eliminate possible errors that commonly occur during sample collection, handling, preservation, and storage. Current research focuses on improving the collection of soils contaminated with volatile organic compounds including an examination of the influence of sample size on "representativeness" of VOC results, VOC releases due to sample disturbance, and the penetrability of caliche to NAPLs.					
Office of Research and Development	National Exposure Research Laboratory (Research Triangle Park, N.C.)	Analytical Methods for Ground Water and Soils	\$1.7M	Intra	http://www.epa.gov/crdlvweb/
Research in the application of advanced instrumentation to soils and groundwater characterization focuses on methods that will provide high-quality data rapidly with simple and rugged protocols. Emphasis is on technologies that can be used to perform analysis in the field, those that can determine pollutants that are intractable by conventional methods, and those that improve risk assessments by providing specific information on the most hazardous forms of pollutants. Currently, pollutants of primary interest are polycyclic aromatic hydrocarbons, chlorinated organics, petroleum-related compounds, and toxic metals. Future emphasis will be on innovative methods and technologies to evaluate and characterize the natural attenuation of contaminants in groundwater and soils.					

Office of Research and Development	National Center for Environmental Assessment (Washington, D.C.; Research Triangle Park, N.C.; and Cincinnati, Oh.)	Superfund Health Risk Assessment	$2.1M	Intra	http://www.epa.gov/ncea/

This research develops methodologies, models, and factors that will enable risk assessors to derive an accurate estimate of the amount of a contaminant found in the soil that is biologically "available" to humans. Future emphasis will include developing statistical distributions for exposure factors to facilitate probabilistic analysis; further refining and validating the biokinetic model for lead; developing better models and methods for the dermal route of exposure; and evaluating the bioavailability of soil-borne contaminants. In addition, research is conducted to develop models and factors to predict the relative toxicity of complex mixtures of ground water contaminants compared to their individual toxicities. This research attempts to answer the question of whether mixtures of ground water contaminants produce a more toxic response, a less toxic response, or no net change in human response.

Office of Research and Development	National Center for Environmental Assessment (Washington, D.C.; Research Triangle Park, N.C.; and Cincinnati, Oh.)	Superfund Ecological Risk Assessment	$1.0M	Intra	http://www.epa.gov/ncea/

This research develops methodologies and factors that will enable ecological risk assessors to estimate the amount of soil-borne contamination that is biologically "available" to wildlife. The major area of emphasis will be to develop ecological soil screening values for common soil contaminants.

Office of Research and Development	National Risk Management Research Laboratory (Cincinnati, Oh.)	Ground Water Remediation	$4.2M	Intra	Http://www.epa.gov/ ada/kerrcenter.html

This research addresses priority remediation problems for ground water and major subsurface sources of ground water contamination, such as NAPLs. Research includes treatment, containment and related source characterization, and ground water fate (including natural attenuation) and transport modeling. DNAPLs are a major source of ground water contaminate for which there are few effective commercialized remediation options. Research on dissolved organic contaminants will continue and will include studies of the application and long-term performance of permeable reactive barriers (PRBs) to chlorinated organics. Research on improved indicators for monitored natural attenuation (MNA) of organics will be continued. Research will be expanded on the remediation of dissolved inorganic plumes and related source areas. Research on MNA of dissolved metals will shift from laboratory to field studies, as will studies of the application of PRBs to arsenic. Research on biotreatment of metal contaminants will continue. Work will be initiated on the application of PRB concepts to metal contaminants in source areas.

TABLE 4.1 Continued

(Note: Funding information is for fiscal year 1999 unless otherwise noted. The funding levels are for comparison purposes only and are approximate figures based on the best available data. In some cases reliable data could not be acquired. This list is not comprehensive.)

Section, Department, or Division	Program	Sub-program	Funding Level	Intra- or Extramural	Web Site for Source and More Information
Office of Research and Development	National Risk Management Research Laboratory (Cincinnati, Oh.)	Soil Remediation	$5.6M	Intra	http://www.epa.gov/ORD/NRMRL/lrpcd/
This research evaluates and develops more cost-effective techniques for cleanup of priority contamination problems in soils, the vadose zone, and sediments. The program will expand research on sediments contaminated with persistent organics and metals. Bench- and field-scale studies will be conducted to develop and refine methods to assess MNA effectiveness. Emerging remediation techniques for in situ sediments will be evaluated, along with improved methods for assessing their performance. Studies will be conducted of the effectiveness of the biotreatment of dredged sediments to allow their beneficial use. Research on the immobilization of metals in soils will continue, focusing on completing lead studies and continuing cadmium studies. Concurrent testing will be done on changes in metal bioavailability as a result of immobilization.					
Office of Research and Development	National Risk Management Research Laboratory (Cincinnati, Oh.)	Containment	$1.9M	Intra	http://www.epa.gov/ORD/NRMRL/lrpcd/
This research addresses the effectiveness of current containment systems and developing new systems using innovative materials and methods. The study includes caps, covers, and vertical barriers for the vadose zone, as well as fixed barriers and pumping methods for containing contaminated plumes. Research on barriers will address the long-term maintenance, effectiveness, and materials that could make them more effective.					
Office of Research and Development	National Center for Environmental Research and Quality Assurance (NCERQA) (Washington, D.C.)	Hazardous Substance Research Centers/ Minority Centers	$8.9M	Both	http://es.epa.gov/ncerqa/hsrc.html http://es.epa.gov/ncerqa/mibc.html
The mission of the Hazardous Substance Research Centers Program is to conduct research to develop and demonstrate new methods that assess and remediate sites contaminated with hazardous substances; improve existing treatment technologies; decrease the production and use of hazardous substances; and educate hazardous substance management professionals and improve community public awareness. Five multi-university centers, each located in a pair of EPA regions, focus on different aspects of hazardous substance management, bringing together researchers from a variety of disciplines to collaborate on integrated research projects. Additionally, two targeted centers serve the particular research needs of minority and disadvantaged communities. These centers address such issues as environmental justice, risk communication and perception of risk, correlation with socioeconomic status, and exposure to environmental contaminants.[h]					

Office of Research and Development	NCERQA (Washington, D.C.)	Joint Program on Bioremediation (EPA is lead agency; DOE, EPA, NSF, and ONR are also involved)	$1.0M	Extra	http://www.epa.gov/docs/ordntrnt/OER-Web/grants96/joint/index.html http://es.epa.gov/ncerqa/rfa/bio.htmll

This program focuses on research that aims to further fundamental understanding of the chemical, physical, and biological processes influencing the bioavailability and release of chemicals in soil, sediments, and ground water under natural conditions and also the role of chemical contaminants that, when released from the medium and assimilated by living organisms, result in adverse effects. The research objective should be to understand the commonality of processes and/or environmental effects involved in contaminant release, movement, and assimilation in order to determine broadly applicable techniques for measuring the potential impacts in complex matrices. Mechanistic and kinetic studies are encouraged. These data can then be used to construct models that increase our understanding of bioavailability (Note: There are no new efforts being funded under this grant, but not all activities and reports under the grant have been completed.)

Office of Research and Development	NCERQA (Joint Program with NIEHS)	Complex Mixtures	$2.7M	Extra	http://es.epa.gov/ncerqa/rfa/chem.html

The goals and scope of this program are to encourage and support research on chemical mixtures that will take advantage of the latest advances in computational and information technologies and molecular biology techniques. The focus is on the mechanistic basis for chemical interactions in biological systems and related health effects and development of better mathematical tools for risk assessment. Since there is a general lack of knowledge concerning the characterization of real-life mixtures based on human exposure and body burden, research will be supported that focuses on exposure assessment, including environmental transport and fate.

Office of Research and Development	NCERQA	Science to Achieve Results (STAR) Program/ Environmental Research Grants	Approximately $100M annually (funding levels vary)	Extra	http://es.epa.gov/ncerqa/grants/ http://es.epa.gov/ncerqa/rfa/

EPA's research programs seek to reduce risks to public health and ecosystems and the uncertainty associated of environmental health risk assessment and management, according the highest priority to those areas where uncertainty is high and that are in critical need of new concepts, methods, and data. EPA also fosters the development and evaluation of new risk reduction technologies, including pollution prevention, end-of-pipe controls, remediation, and monitoring. EPA's extramural research grant programs are administered by the National Center for Environmental Research and Quality Assurance through the Science to Achieve Results Program, which has four formal solicitation periods during the year—January, April, August, and October. Requests for Applications invite research proposals from academic and non-profit institutions located in the U.S., and state and local governments. Research topics vary. The 1999 solicitations can be found on http://es.epa.gov/ncerqa/grants/99grants.html.

TABLE 4.1 Continued

(Note: Funding information is for fiscal year 1999 unless otherwise noted. The funding levels are for comparison purposes only and are approximate figures based on the best available data. In some cases reliable data could not be acquired. This list is not comprehensive.)

Section, Department, or Division	Program	Sub-program	Funding Level	Intra- or Extramural	Web Site for Source and More Information
U.S. DEPARTMENT OF HEALTH AND HUMAN SERVICES					
National Institutes of Health	National Institute of Environmental Health Sciences	Superfund Hazardous Substances Basic Research Program (this is a joint program between NIEHS and EPA)	Approximately $37M in FY 2000 to fund 15 to 20 new grants or competitive continuation grants	Extra	http://www.niehs.nih.gov/sbrp/newweb/sbrpres.htm
Research efforts are being pursued that are focused on (1) understanding the risks hazardous waste exposures pose to humans, and (2) developing new technology that will help remediate contaminated sites. This program was established by the Superfund Amendments and Reauthorization Act of 1986. It receives its funding through an interagency agreement with the EPA using Superfund Trust monies and is administered by the National Institue of Environmental Health Sciences.					
NATIONAL SCIENCE FOUNDATION					
Biological Sciences	Environmental Biology Division	(Including Long Term Ecological Research Sites Program)	FY 98: $79.3M FY 99: $85.9M	Extra	http://www.nsf.gov/bfa/bud/fy1999/start.htm http://www.nsf.gov/bfa/bud/fy2000/ http://lternet.edu/index.html
The Division of Environmental Biology supports fundamental research on the origins, functions, relationships, interactions, and evolutionary history of populations, species, communities, and ecosystems. Scientific emphases include biodiversity, molecular evolution, mesoscale ecology, computational biology (including modeling), conservation biology, global change, and restoration ecology. The division also supports a network of long-term ecological research sites;[i] doctoral dissertation research; research conferences and workshops; undergraduate mentoring in environmental biology; and a variety of foundation-wide activities.					

Biological Sciences	Plant Genome Research		FY 98: $40M FY 99 $50M	Extra	http://www.nsf.gov/bfa/bud/fy1999/start.htm

The plant genome research subactivity, begun in FY 1998, supports research to advance understanding of the structure, organization, and function of plant genomes. This effort is built on an existing base of genome research supported throughout the biological sciences division. Enhanced support for fundamental research will accelerate the use of new knowledge and innovative technologies toward a more complete understanding of basic biological processes in plants, with emphasis on such economically significant species as corn.

Engineering	Bioengineering and Environmental Systems	Environmental Engineering Program	FY 98: $28.8M FY 99 $32.42M (this is for all of Bioengineering and Environmental Systems)	Extra	http://www.eng.nsf.gov/bes/Programs/Environmental_Engineering_Basi/environmental_engineering_basi.htm

The Environmental Engineering Program supports sustainable development research with the goal of applying engineering principles to reduce adverse effects of solid, liquid, and gaseous discharges into land, fresh and ocean waters, and air that result from human activity and impair the value of those resources. This program also supports research on innovative biological, chemical, and physical processes used alone or as components of engineered systems to restore the usefulness of polluted land, water, and air resources. Engineering principles underlying pollution avoidance, as well as pollution treatment and reparation are emphasized. Improved sensors, innovative production processes, waste reduction and recycling, and industrial ecology are important to this program. Research may be directed toward improving the cost effectiveness of pollution avoidance, as well as developing fresh principles for pollution avoidance technologies.

Engineering	Chemical and Transport Systems (CTS)	Fluid and Particle Processes Program	FY 98: $39.67M FY 99: $42.14M (for all of Chemical and Transport Systems)	Extra	http://www.eng.nsf.gov/cts/html/fluid.htm

This program supports fundamental and applied research on mechanisms and phenomena governing single and multiphase fluid flow, particle formation and transport, various multiphase processes, nanostructures, and fluid-solid system interaction. Research is sought that contributes to improving basic understanding and design, predictability, efficiency, and control of existing systems that involve dynamics of fluids and particulates; and the innovative uses of fluids and particulates in materials development, manufacturing, biotechnology, and the environment.

Engineering	Chemical and Transport Systems (CTS)	Interfacial, Transport, and Separation Process	(see above)	Extra	http://www.eng.nsf.gov/cts/html/interfacial.htm

This program supports research in areas related to interfacial phenomena, mass transport phenomena, separation science, and phase equilibrium thermodynamics. Research in these areas supports various aspects of engineering technology with major impact on chemical and material processing, as well as bioprocess engineering. Research in this program also contributes to the division emphasis on basic knowledge impacting on physicochemical hazardous waste treatment and avoidance. The program provides support for new theories and approaches determining the thermodynamic properties of fluids and fluid mixtures in biological and other fluids with complex molecules. Separations research is directed at many areas with special emphasis on bioprocessing and all forms of chromatographic, membrane, and special affinity separations.

TABLE 4.1 Continued

(Note: Funding information is for fiscal year 1999 unless otherwise noted. The funding levels are for comparison purposes only and are approximate figures based on the best available data. In some cases reliable data could not be acquired. This list is not comprehensive.)

Section, Department, or Division	Program	Sub-program	Funding Level	Intra- or Extramural	Web Site for Source and More Information
Engineering	Civil and Mechanical Systems	Construction/ Geotechnology/ Structures	FY 98: $44.71M; FY 99 (requested): $59.5M (for all of Civil and Mechanical Systems)	Extra	http://www.eng.nsf.gov/ cms/CGS/cgs.htm
Research is supported that increases geotechnical knowledge of foundations, slopes, excavations, and other geostructures, including soil and rock improvement technologies and reinforcement systems; constitutive modeling and verification in geomechanics; remediation and containment of geoenvironmental contamination; transferability of laboratory results to field scale; and nondestructive and in situ evaluation. Research is also supported that will advance the knowledge base on advanced polymer materials; high performance steel and concrete materials; durability of construction materials; safety and reliability of bridges, including condition assessment; and indoor environmental conditions, such as air quality and personnel comfort in buildings. Also of interest are activities that will increase the present understanding of the science and technology used to design, analyze, diagnose, repair, remediate, retrofit, and enhance the performance of constructed facilities and interactions between natural and constructed environments, and to use knowledge gained to improve the management and performance of new and existing infrastructure systems and facilities.					
Mathematical and Physical Sciences	Chemistry	Inorganic, Bioinorganic, and Organometallic	FY 98: $113M FY 99 request: $125.3M (for all Chemistry)	Extra	http://www.nsf.gov/ mps/chem/prgminfo/programs.htm
This program supports research on synthesis, structure, and reaction mechanisms of molecules containing metals, metalloids, and nonmetals encompassing the entire periodic table of the elements. Included are studies of stoichiometric and homogeneous catalytic chemical reaction; bioinorganic and organometallic reagents and reaction; and the synthesis of new inorganic substances with predictable chemical, physical, and biological properties. Such research provides the basis for understanding the function of metal ions in biological systems, for understanding the synthesis of new inorganic materials and new industrial catalysts, and for systematic understanding of chemistry of most of the elements in the environment.					
Mathematical and Physical Sciences	Chemistry	Organic Chemical Dynamics	(see above)	Extra	http://www.nsf.gov/ mps/chem/prgminfo/programs.htm
This program supports research on the structures and reaction dynamics of carbon-based molecules, metallo-organic systems, and organized molecular assemblies. Research includes studies of reactivity, reaction mechanisms, reactive intermediates, and characterization and investigation of new organic materials. Such research provides the basis for understanding and modeling biological processes and for developing new or improved theories relating chemical structures and properties.					

Geosciences	Earth Sciences	Hydrologic Sciences	FY 98: $58.6M FY 99 request: $65.7M (for all of Earth Sciences)	Extra	http://www.geo.nsf.gov/ear/earcore.htm

The Hydrologic Sciences Program supports basic research dealing with the earth's hydrologic cycle and the role of water on and near the continental surfaces. The Program views hydrologic sciences as a geoscience interactive on a wide range of space and time scales with the ocean, atmospheric, and solid earth sciences as well as plant and animal sciences. Supported projects may involve water in the form of precipitation, lakes, streams, and ground water, and interactions with landforms, soils, the atmosphere, the biosphere, and the Earth's crust. The program encourages integrated studies of water balance and fluxes among the various reservoirs.

Geosciences	Earth Sciences	Geophysics	(see above)	Extra	http://www.geo.nsf.gov/ear/earcore.htm

The Geophysics Program supports laboratory, field, theoretical, and computational studies related to composition, structure, and processes of the Earth's interior. Topics include studies in seismicity and seismic wave propagation; the nature and occurrence of earthquakes; magnetic, gravity, and electrical fields; and internal temperature distribution. Supported research also includes geophysical studies of active deformation, including GPS-based geodesy, and fundamental laboratory studies of properties and behavior of Earth materials in support of geophysical observation and theory.

Geosciences	Earth Sciences	Geology and Paleontology	(see above)	Extra	http://www.geo.nsf.gov/ear/earcore.htm

The Geology and Paleontology Program supports studies of physical, chemical, geological, and biological processes at or near Earth's surface and the landforms, sediments, fossils, low-temperature fluids, and sedimentary rocks that they produce. Topics represented in the program include paleontology, paleoecology, stratigraphy, paleoclimatology, geomorphology, glacial geology, sedimentology, soil genesis, sedimentary petrology, diagenesis, and organic geochemistry.

Geosciences	Crosscutting Programs	Environmental Geochemistry and Biogeochemistry	Approximately $4.8M	Extra	http://www.nsf.gov/home/crssprgm/egb/start.html

The goal of the Environmental Geochemistry and Biogeochemistry activity is to enhance fundamental interdisciplinary research on chemical processes that determine the behavior and distribution of inorganic and organic materials in environments near Earth's surface. Of particular importance are projects that characterize chemical parameters in both perturbed and unperturbed natural systems, clarify the chemical and biological processes or behavior observed, or combine observations and interpretations into predictive models. No new awards are planned after the end of FY99.

TABLE 4.1 Continued

(Note: Funding information is for fiscal year 1999 unless otherwise noted. The funding levels are for comparison purposes only and are approximate figures based on the best available data. In some cases reliable data could not be acquired. This list is not comprehensive.)

Section, Department, or Division	Program	Sub-program	Funding Level	Intra- or Extramural	Web Site for Source and More Information
NUCLEAR REGULATORY COMMISSION					
Office of Nuclear Regulatory Research	Radiation Protection, Environmental Risk, and Waste Management Branch		(Funding levels are listed for individual projects below and are for planning purposes only.)	Both	Web page describing all on-going Office of Research projects is under development
The research conducted in this branch is designed to maintain or improve tools used to evaluate the consequences of decisions regarding the disposition of sites and/or facilities contaminated with radioactive material. This work is a combination of applied and basic research. The basic research is directed at better understanding and modeling processes, systems, and events that may effect simulations of the evolution of contaminated systems over time and the accompanying potential exposures of the public to low levels of radiation. The applied research is directed toward incorporating these process, system, and event models into user-friendly computational tools.					
Office of Nuclear Regulatory Research	Radiation Protection, Environmental Risk, and Waste Management Branch	Engineered Barrier Performance	$100K	Both	
This program is developing models to assess the long-term performance of engineered barriers. A concrete model has been developed and is being validated through the collection and study of archeological samples. Future work is planned to look at long-term degradation of non-concrete engineered barrier materials. The work is conducted in cooperation with the National Institute of Science and Technology.					
Office of Nuclear Regulatory Research	Radiation Protection, Environmental Risk, and Waste Management Branch	Source Term Characterization	$100K	Intra	
This work is being conducted to establish the mechanisms of degradation and release rates for radioactive materials from mineral phases found in slags from ore processing. Many of the same radionuclide-bearing mineral phases will be found in artificial waste forms such as borosilicate glass. Contemporary and archeological slags are being examined with sophisticated analytical equipment at The Johns Hopkins University to identify radionuclide bearing mineral phases and then establish degradation rates for periods of up to a thousand years. This work will end in March 2000.					

Office of Nuclear Regulatory Research	Radiation Protection, Environmental Risk, and Waste Management Branch	Sorption Mechanisms	$550K (in two separate projects)	Extra

Sorption processes that control the retardation of radioactive materials moving in ground water are being studied to provide better models for this complex process. This work is being conducted by Sandia National Laboratories (low to mid atomic number radionuclides) and the U.S. Geological Survey (uranium and its decay daughters).

Office of Nuclear Regulatory Research	Radiation Protection, Environmental Risk, and Waste Management Branch	Conceptual Model Uncertainty	$400K	Extra

Traditional estimates of uncertainty and variability focus on the heterogeneity of natural systems and the parameters that describe those systems as well as the accuracy of measurement techniques. Data used to support models are often subject to multiple interpretations but the uncertainty associated with selecting one interpretation for modeling purposes is seldom addressed or quantified. This project is systematically looking at the process of data collection and model selection for hydrogeologic systems and will provide a methodology for addressing the uncertainty associated with alternative conceptual models. This work is being conducted at the University of Arizona and will extend into FY 2002.

Office of Nuclear Regulatory Research	Radiation Protection, Environmental Risk, and Waste Management Branch	Deterministic Effects of Occupational Doses	$300K	Extra

Radiation exposure data and medical histories for workers at the Mayak Production Association in Russia are being analyzed under a joint project with Russian scientists to establish a basis for estimating the effects of such exposures on humans. The work is being conducted at the University of Pittsburgh and will extend through FY 2002.

Office of Nuclear Regulatory Research	Radiation Protection, Environmental Risk, and Waste Management Branch	Radionuclide Solubilities	$150K	Extra

Work being conducted to extend the data base on the solubilities of radioactive species under different environmental conditions such as pH, Eh, and ionic strength. Work is being conducted at Pacific Northwest National Laboratory and is planned to continue through FY 2000.

TABLE 4.1 Continued

(Note: Funding information is for fiscal year 1999 unless otherwise noted. The funding levels are for comparison purposes only and are approximate figures based on the best available data. In some cases reliable data could not be acquired. This list is not comprehensive.)

Section, Department, or Division	Program	Sub-program	Funding Level	Intra- or Extramural	Web Site for Source and More Information
Office of Nuclear Regulatory Research	Radiation Protection, Environmental Risk, and Waste Management Branch	Parameter Uncertainty	$600K	Extra	
Three projects are looking at default parameter values and assumptions for performance assessment models. Both hydrogeologic parameters and health effects parameters are being examined to identify and document the basis for current parameters or recommend new parameters and an associated technical basis. This work is being carried out at Pacific Northwest National Laboratory, Argonne National Laboratory, and Sandia National Laboratories and will continue through FY 2000.					
Office of Nuclear Regulatory Research	Radiation Protection, Environmental Risk, and Waste Management Branch	Performance Assessment Code Development and Maintenance	$450K	Extra	
A flexible, integrated model capable of assessing the performance of a complex contaminated site or disposal facility is being developed. The model is being developed with a modular structure to allow various process models to be included and within a fully probabilistic framework. This work is being conducted by Sandia National Laboratories and the current phase will continue through FY 2002.					

[a] About a quarter of this is directed toward site characterization and fate and transport research.

[b] Cooperative support—research done by Center for Multiphase Research, Department of Environmental Sciences and Engineering, University of North Carolina.

[c] WES Fate and Effects program is both direct- and reimbursably-funded. Headquarters, U.S. Army Corps of Engineers, provides the direct funds and Army installations may or may not provide the reimbursable funds.

[d] This laboratory operates as a Navy Working Capital Fund activity. As such, all costs, including overhead, must be recovered from various sponsors. Funding comes from the Chief of Naval Research, the Naval Systems Commands, and other government agencies, such as the U.S. Air Force, the Advanced Research Projects Agency, the Department of Energy, and the National Aeronautics and Space Administration, as well as several nongovernment entities.

[e] Includes facility operations. FY 99 numbers are for research only: Engineering: $17.471M, Geosciences: $24.189M.

[f] These are total program costs including facility operations. FY98 "scientific research" $187M.

[g] The National User Facilities are maintained by both of these entities. These user facilities are listed in Table 4.2

[h] The Hazardous Substances Research Centers are listed in the text.

[i] The NSF provides partial support for 21 Long Term Ecological Research Sites. The Division on Biology provides support for 19 of these 21. In FY 99 $12.7M of biology funding was allocated for this program.

5

Knowledge Gaps and Research Needs

The statement of task for this study directed the committee to identify significant knowledge gaps relevant to subsurface contamination problems at DOE sites and to provide recommendations for a long-term basic research program to fill those gaps (see Sidebar 1.1). In this chapter, the committee identifies what it judges to be the significant knowledge gaps that emerged from its review of DOE's subsurface contamination problems in Chapter 2 and, for each identified gap, the committee provides a short discussion of basic research needs. This information will be used to formulate recommendations for a long-term research program in Chapter 6.

For purposes of this discussion, the committee defines "knowledge gap" as a deficiency in scientific or engineering understanding that is now, or likely will be in the future, a significant impediment to DOE's efforts to complete its mission to clean up, stabilize, or contain subsurface contamination. Perhaps the most direct manifestation of a DOE knowledge gap is a technology gap, that is, a deficiency in technical capabilities to identify and deal with contamination problems. The committee has not focused on technology gaps in this report; that is the topic of another recent NRC report (NRC, 1999). Rather, the committee has focused on the identification of the knowledge gaps that underpin those technology gaps.

The committee has been selective in the identification of subsurface contamination knowledge gaps and research needs for the EM Science Program. The identification of knowledge gaps involves an appreciable element of judgment on the part of the committee, especially in interpreting the significance of the subsurface contamination problems (see Chapter 2) and the scope and objective of other federal research programs. The committee believes that the gaps it has identified are highly significant and that they must be addressed through basic research if the DOE cleanup program is to succeed. Further, the committee believes that a focus on these knowledge gaps is likely to yield the greatest payoffs for DOE in terms of enhanced cleanup capabilities at reduced costs and risks at major DOE sites.[1] This is especially true given the small size of the EM Science Program relative to the scope of the DOE cleanup mission. The annual budget for the EM Science Program budget is on the order of $30 million to $50 million and is used to support basic research related to all aspects of the cleanup mission. This is less than 0.1 percent of the total EM annual budget of $5.8 billion. Without a significant increase in its budget, the EM Science program is unlikely to have a significant impact on DOE cleanup effectiveness and costs.

Organizing Scheme Used in This Analysis

The committee identified significant knowledge gaps and research needs through discussions and analyses of the "snapshot" of DOE's subsurface contamination problems presented in Chapter 2. To organize this analysis and ensure its completeness, the committee developed the organizing scheme shown in Figure 5.1. This organizing scheme is based partly on the approach used by the Subsurface Contaminants Focus Area to organize its technology development programs (see Figure 3.2), but it also includes the data collection and analysis steps that provide the supporting information needed to make appropriate corrective action decisions.[2] The committee's organizing scheme, hereafter referred to as the framework for site remediation, is described briefly in the following paragraphs.

---

[1]As discussed in Chapter 2, the major sites represent DOE's largest future mortgages and longest-term commitments.

[2]The committee uses the term "corrective action" in the following discussion to refer to actions taken by DOE to address its subsurface contamination problems. A corrective action can range from no action in cases where the subsurface contamination is thought to pose minimal hazards to humans or the environment, or where remediation is infeasible, to aggressive actions to treat, remove, or contain contamination that poses significant hazards. As noted in Chapter 1, the term is sometimes used interchangeably with terms like "cleanup" and "remedial action," but it really encompasses a broader range of possible options for dealing with contamination, because it includes the no action (i.e., no cleanup or no remedial action) option.

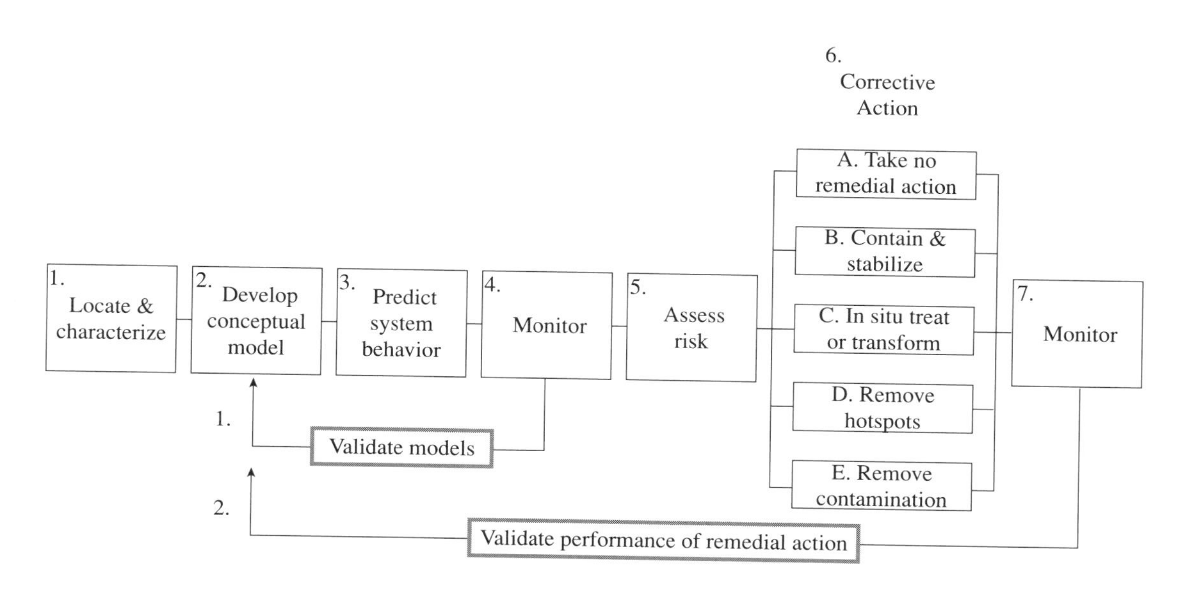

FIGURE 5.1 Framework for site remediation.

The boxes in Figure 5.1 represent each of the major steps in the remediation process, and the arrows represent decision and assessment points. Boxes 1 through 5 represent the process that could be followed to develop information to make an appropriate corrective action decision. The initial step is focused on locating and characterizing the contaminants of concern (Box 1). This step involves determining the spatial distributions, types, amounts, and physical and chemical states of subsurface contaminants, as well as the subsurface properties that affect contaminant fate and transport behavior. Locating and characterizing contamination in the subsurface may be done using direct (e.g., drilling and sampling) and indirect (e.g., surface and borehole geophysical) techniques.

The location and characterization data obtained in the first step are then used to develop a conceptual model of the site (Box 2), that is, a description of the subsurface as estimated from knowledge of the known site geology and hydrology and the physical, chemical, and biological processes that govern contaminant behavior. The conceptual model provides a descriptive framework for assessing how the subsurface system will behave with passing time and in response to potential corrective actions. As noted later in this discussion, the conceptual model is improved over time as more information on subsurface conditions and processes becomes available.

The conceptual model provides a basis for constructing more quantitative dynamic models that can be used to predict behavior (Box 3) of the subsurface system over a specified time period. These predictive models are developed from mathematical representations of the conceptual model. In current practice, most predictive models are discrete representations of the physically continuous subsurface system and are typically solved with such numerical techniques as finite elements and finite difference.[3] The parameters in these models represent the physical, chemical, and biological characteristics at each point in the subsurface. Parameters of interest in predictive models quantify the relationship between the driving forces (e.g., hydraulic gradient and chemical concentration gradient) and the resulting behavior (e.g., flow and transport). In the case of discrete models, the parameter values are meant to represent volume-averaged properties around each modeled point.

The predicted system behavior is then compared to the observed behavior as measured in the field through monitoring activities (Box 4). A feedback loop (Arrow 1) updates the conceptual and predictive models when the behaviors do not match according to some specified measure(s) of comparison. The process of testing the predictive model to determine whether it appropriately represents the system behavior of interest is usually referred to as model validation.[4] Because of uncertainties in the model and data any match between predicted and observed behaviors is only possible in a statistical sense. Consequently, validation is best thought of as a process of confidence building through increased understanding of fundamental mechanisms in the underlying system rather than as a process to confirm or prove the correctness of a model.

The predictive model can be used to understand the present behavior of the subsurface system and to estimate future contaminant migration to assess risk to human and environmental health (Box 5). A corrective action decision (Box 6A-E) that will reduce risk to acceptable levels is then made using the information developed in the risk assessment. The corrective action can range from no action (Box 6A) to remove contamination (Box 6E).

During and following the corrective action, monitoring activities (Box 7) are again employed to assess the efficacy of that action. Long-term monitoring is usually required to confirm the effectiveness of, or to gain regulatory approval for, approaches that involve no action or containment and stabilization. Inconsistencies between the measured and

---

[3]Continuous representations are sometimes used in analytical models for screening-level assessments.

[4]In model development protocol, a step referred to as model verification involves evaluating whether the numerical model solves the mathematical equations of the conceptual model with acceptable accuracy. In this discussion, model verification is included as a step in the validation process.

predicted performance of the corrective action may indicate that the conceptual model of the system is deficient or that the parameters of the model are not sufficiently resolved, and it may be extremely difficult to know which is the case. In these cases, there is a feedback loop (Arrow 2) to the conceptual model (Box 2) through the predictive model, which must be updated so that the corrective action decision process can be revisited.

Although the framework for site remediation shown in Figure 5.1 is presented as a linear process, it is in reality an observational procedure that follows both parallel and iterative paths. The framework may be traversed many times as new information is acquired and incorporated into the conceptual and predictive models and as corrective action performance is assessed.

There is some correspondence between the organizing scheme outlined in Figure 5.1 and the technology development organizing scheme shown in Figure 3.2. For example, the identify function in Figure 3.2 is roughly equivalent to the locate and characterize function in Figure 5.1. Similarly, the validate function in Figure 3.2 is roughly equivalent to the validate performance of remedial action function (Arrow 2) in Figure 5.1. The remaining functions in Figure 3.2 have no directly equivalent functions in Figure 5.1, and there are many functions in Figure 5.1 that are not represented at all in Figure 3.2 (e.g., the develop conceptual model and predict system behavior functions). The organizing scheme shown in Figure 5.1 is more complete than that given in Figure 3.2.

Knowledge Gaps

The committee identified significant knowledge gaps in the following process steps in the framework for site remediation shown in Figure 5.1:

- location and characterization of subsurface contaminants and characterization of the subsurface (Box 1);
- conceptual modeling (Box 2);
- containment and stabilization (Box 6B); and
- monitoring and validation (Boxes 4 and 7 and Arrows 1 and 2).

These knowledge gaps do not include those associated with active remediation of subsurface contamination (Boxes 6C-6E), with the exception of remediation monitoring. This may come as a surprise to some readers, given that the current EM Science Program portfolio is heavily focused on this area (see the closing section of Chapter 3). The committee did not highlight knowledge gaps on these process steps, because subsurface contamination is highly distributed at many DOE sites, making cost-effective remediation infeasible, and because EM Science Program resources are limited and there is much work on these topics in other federal research programs (see Chapter 4).

Location and Characterization of Subsurface Contaminants and Characterization of the Subsurface

An important conclusion that emerges from the committee's analysis of subsurface contamination problems in Chapter 2 is that the capabilities to locate and characterize subsurface contaminants at many DOE sites are incomplete. This conclusion is perhaps best supported by the following three examples from Chapter 2:

- subsurface radionuclide contamination in the 200 Area at Hanford;
- mixed contaminant plumes at Test Area North at the Idaho Site; and
- contaminant plumes and hot spots in waste burial grounds at the Savannah River Site.

Locating contamination in the subsurface at DOE sites has focused on three interrelated approaches: (1) information derived from historical operations and records; (2) direct observations of contamination on the surface, in surface water, and in boreholes; and (3) indirect geochemical and geophysical measurements from the surface and in boreholes. These three sources of information have been used at some sites to develop predictive models of contaminant movement in the subsurface, and the predictive models have been tested by further direct observations and measurements. Frequently, these models have not captured the essential behavior of the contaminant, either in direction or speed of movement.

The challenges of locating subsurface contamination are magnified by the wide range of contaminant types (e.g., mixtures of organic solvents, metals, and radionuclides) in the subsurface at many DOE sites (see Chapter 2); the wide variety of geological and hydrological conditions across the DOE complex (see Table 2.2); and the wide range of spatial resolutions at which this contamination must be located and characterized, ranging from widely dispersed contamination in groundwater plumes to small isolated hot spots in waste burial grounds. Moreover, because contaminant migration involves dynamic transport processes, continuous temporal information on contaminant locations is required. In effect, location, characterization, and continuing monitoring efforts must be integrated to assure an adequate database for planning and implementing appropriate corrective actions.

Fundamental advances in capabilities to locate and characterize subsurface contamination and important subsurface properties will help DOE better assess the potential hazards of its contamination problems and to design and implement appropriate corrective action strategies (e.g., see Sidebar 5.1). Moreover, research on subsurface heterogeneity in geology, geochemistry, hydrology, and microbiology will provide a framework for assessing the fate and transport of contaminants. Examples of significant knowledge gaps include the following:

- *Locating contaminants in the subsurface.* At many sites, the points of entry of contaminants into the subsurface (e.g., through a leaking waste burial ground or injection well) are at least approximately known. However, the determination of the spatial distributions of contaminants (that may or may not change with time) once they enter the subsurface remains a major knowledge gap. Currently available indirect measurement methods (e.g., geophysical methods) are inadequate for locating most types of contaminants in the subsurface, and direct methods such as drilling are both expensive and limited in effectiveness, because they only provide samples from specific points in the subsurface along the borehole. Moreover, boreholes provide potential contaminant transport pathways through the subsurface.
- *Characterizing contaminants in the subsurface.* Once contaminants enter the subsurface, they can act as long-term sources of pollution to ground or surface water. Understanding how to characterize the concentrations, speciations, and release rates of contaminants in the subsurface is a significant knowledge gap across the DOE complex. In general, there are poor records of contaminant discharges to the subsurface, so contaminant quantities are highly uncertain. Moreover, once contaminants enter the subsurface they can move long distances, either diffusing through the fluid medium or migrating as a distinct plume, leading to contaminant distributions that are variable in size, shape, and concentration. Currently available direct and indirect observing technologies[5] have limited effectiveness for character-

---

[5]Direct observing technologies allow in situ measurements or samples to be obtained (e.g., by drilling). Indirect observing technologies allow measurements to be made remotely (e.g., through geophysical measurements of the subsurface). The terms "invasive" and "noninvasive" are sometimes used synonymously, but this usage is not strictly correct. Indirect measurements can be obtained through inva-

izing site conditions and defining the extent and concentrations of contaminant bodies.

- *Characterizing physical, chemical, and biological properties of the subsurface, including improved approaches to understanding the properties of the geologic system and relating them to contaminant fate and transport.* The subsurface characteristics at a site place fundamental controls on contaminant fate and transport behavior. Subsurface characteristics also govern the selection of conceptual and predictive models as well as the application and effectiveness of appropriate corrective actions. The knowledge gaps include understanding which characteristics control fate and transport behavior in the subsurface and also understanding how those characteristics can be measured at the appropriate scales over large subsurface volumes, using both indirect and direct techniques. The integration of direct measurements of subsurface geologic properties with indirect measurements (e.g., from geophysical methods) has been used very successfully in the petroleum industry to develop conceptual and quantitative models of subsurface transport. Such methods are potentially applicable to DOE sites.
- *Characterizing highly heterogeneous systems.* This knowledge gap is a special case of the previous knowledge gap and is a significant problem at many DOE sites, which are very large in spatial extent and exhibit intra- and inter-site variations in geologic and hydrologic conditions (see Chapter 2). Heterogeneity arises from the spatial variability in geological, chemical, and biological properties of the subsurface. A fundamental understanding of these properties, and especially the geological framework, is a necessary prerequisite to understanding the fate and transport of contaminants. Heterogeneity may occur at several spatial scales in complex subsurface systems, but they may control contaminant fate and transport processes only at one or a few scales. The primary knowledge gaps are in understanding the heterogeneity scales that govern these processes, how to characterize this heterogeneity without having to perform an exhaustive characterization of the subsurface, and how to represent this heterogeneity in mathematical formulations.

Research Needs

In the committee's judgment, basic research can support the development of new and improved capabilities to locate and characterize

---

sive means, as when borehole geophysical methods are employed to obtain subsurface measurements.

contamination in the subsurface, and also to characterize subsurface properties at the scales that control contaminant fate and transport behavior. Development of the following capabilities is especially needed:

1. Improved capabilities for characterizing the physical, chemical, and biological properties of the subsurface. These approaches should provide information on the following system properties and behaviors at the spatial and temporal scales that control contaminant fate and transport behavior:

    - contaminant locations and characteristics;
    - transport pathways;
    - subsurface properties and boundary conditions that control contaminant fate and transport behavior; and
    - physical, chemical, and biological interactions between contaminants and earth materials.

SIDEBAR 5.1 NEW APPROACHES FOR DIRECT OBSERVING

The major limitations on direct observations by conventional drilling and sampling have been high costs and concerns that direct approaches may unwittingly exacerbate the spread of contaminants in the subsurface. The use of reduced diameter drillholes (using 4- to 6-inch diameter drills) as a cost-saving method has been explored widely in the petroleum industry, but cost reductions have not been encouraging. However, recent developments in miniaturized drilling and sampling technologies (e.g., Albright and Dreesen, 2000) hold promise for significantly reducing drilling costs and reducing the potential for contaminant spread when these technologies are used at DOE sites.

A new technology, microdrilling, represents the kinds of advanced capabilities made possible by basic scientific and engineering research. This technology uses coiled tubing, steerable miniature-diameter (1 3/8 inches to 2 inches [3.5 centimeters to 5.1 centimeters]) down-hole motors, and down-hole microinstrumentation to obtain in situ measurements and samples of contaminated subsurface environments. Additionally, smaller diameter holes reduce contaminant migration potential and promote more effective sealing.

The depth capabilities thus far demonstrated are adequate for almost all of the major DOE sites (down to about 300 meters, or about 1,000 feet). Many aspects of this microborehole technology still require extensive research and development, including work on sampling techniques, down-hole instrumentation for diverse measurements, and effective plugging; however, enough feasibility demonstrations have been completed to indicate great promise for use at DOE and other contaminated sites.

Albright and Dreesen (2000) suggest that this technology may cut drilling costs by at least 70 percent compared to conventional technologies. They also suggest that much greater cost savings are possible as these techniques are refined.

Research on indirect observations could involve the development of new approaches for measuring contaminant and subsurface properties (e.g., approaches utilizing "unconventional" geophysical wave attributes such as polarized and nonlinear wave responses) or new ways of interpreting "conventional" observational data to obtain information on the system properties of interest. For direct observations, the research must also address how the observing process changes the system being measured. Approaches for making direct and indirect observations in the unsaturated zone are especially needed.

2. Improved capabilities for characterizing physical, chemical, and biological heterogeneity, especially at the scales that control contaminant fate and transport behavior. Approaches that allow measurements or estimates of heterogeneity features to be obtained directly (i.e., without having to perform a detailed characterization of the subsurface) are especially needed.
3. Improved capabilities for measuring contaminant migration and the system properties that control contaminant movement.
4. Methods to integrate data collected at different spatial and temporal scales to better estimate contaminant and subsurface properties and processes, and also methods to integrate such data into conceptual models.

Conceptual Modeling

As shown by several examples in Chapter 2, DOE is finding subsurface contamination in unexpected places:

- Technetium was discovered in groundwater beneath the SX Tank Farm in the 200 Area at the Hanford Site.
- Plutonium was discovered in colloids in groundwater near the Benham Test at the Nevada Test Site.
- Plutonium was discovered in groundwater beneath the Radioactive Waste Management Complex at the Idaho Site.

These discoveries were "unexpected" because models of the subsurface at these sites did not predict them (e.g., see Sidebar 2.6). Conceptual and predictive models have been developed for subsurface contaminant fate and transport for many DOE sites, but in many cases these models have proven ineffective for understanding and predicting contaminant movement, especially at sites that have thick unsaturated zones or complex subsurface characteristics.

The conceptual model "problem" has many possible causes. The models themselves may be deficient because they were developed using insufficient data on subsurface characteristics, contaminant distributions, or transport processes, or the models may simply have an inappropriate theoretical basis. Good conceptual models must be grounded in sound theory and underpinned with sound and sufficient data. In the committee's judgment, at least part of the problem is that conceptual model development is not viewed as an explicit part of remediation practice. Consequently, there are few standardized tools or accepted methodologies for developing such models, which has led to ad hoc and inconsistent approaches across DOE sites.

Accurate conceptualizations are essential for understanding the long-term fate of contaminants in the subsurface and the selection and application of appropriate corrective actions. The significant knowledge gaps include the following:

- *Contaminant fate and transport.* Understanding the factors controlling the long-term fate of contaminants in the subsurface is important for assessing the potential for human and ecological exposure and for selecting appropriate corrective actions. Understanding the dominant contaminant transport processes and pathways through the subsurface remains a significant knowledge gap for building accurate and useful conceptual and predictive models. The simplest formulation of contaminant transport uses porous media flow of a dissolved phase, but such transport may be the exception at many DOE sites, where transport can occur in several distinct manners (e.g., colloidal transport) through both porous media and fractures and may involve a variety of chemical and biological reactions. The myriad chemical, biological, and physical processes operating in the subsurface operate at different time scales and are poorly understood, especially for metals and radionuclides.
- *Coupling physical, chemical, and biological processes.* The physical, chemical, and biological properties and processes that govern contaminant fate and transport do not act independently. Rather, they interact (i.e., they are coupled) in complex and often poorly understood ways. Many coupled processes operate over very small spatial scales that are defined by a distribution of properties, making it difficult to incorporate representations of these processes into conceptual and mathematical models. For example, redox potential and pH (chemical properties related to bulk mineralogy, biological activity, and fluid composition) can affect either or both physisorption and chemisorption of contaminants onto solid phases. The heterogeneous distribution of permeability (a physical property related to the geological characteristics of the

subsurface) can result in highly variable rates of fluid flow (a physical process). These processes combine to effect transport (a coupled process) of certain metals and radionuclides over small spatial scales. Similarly, the coupling of biomass availability (a property with biological, physical, and chemical components) and substrate availability (controlled by processes such as sorption, dissolution, and transport) with the distribution of electron acceptors (also possessing biological, physical, and chemical controls) can result in spatially variable rates of in situ contaminant biodegradation (a coupled process). The coupling of processes and their control by subsurface properties are only beginning to be understood. Moreover, little progress has been made on how to represent coupled processes in predictive models.

- *Model parameter development.* Model parameters are well understood and definable for very simple homogeneous subsurface systems. However, in highly complex subsurface systems, parameter definition may require unobtainable amounts of detailed characterization data. In these cases, it is important to understand which processes are actually dominating the behavior of the system and to define parameters appropriate to those processes. Determining how to make the appropriate simplifications and approximations is the main thrust of conceptual modeling research that leads to the identification of appropriate model parameters.

The definition and estimation of model parameters requires a good understanding of the subsurface system and transport processes being modeled, which is not often the case at DOE sites. For example, the traditional approach for modeling porous media is to choose permeability as a model parameter. If the porous medium is highly heterogeneous (e.g., if it contains a few large and interconnected fractures) then the generalized concept of permeability is not well defined, and permeability may not be an appropriate characterization of the physical system. Flow and transport may be dominated by the fracture system, and the model parameters should represent the properties of these permeable and connected pathways. Similarly, for fate and transport models, the traditional approach is to assume equilibrium sorption and use the equilibrium partition coefficient as a model parameter. If the sorption reactions are not at equilibrium, however, then the equilibrium partition coefficient by itself is not an appropriate parameter, and additional parameters describing mass transfer kinetics must also be included. The challenge is to define the right conceptualization of the physical, chemical, or biological processes that dominate system behavior, which in turn defines the appropriate model parameters to be used.

The field observations used to develop parameter estimates are made at many different scales and times and provide information about different properties of the subsurface system. Samples from drillhole core, for example, can provide detailed information on the physical, chemical, and biological properties of the subsurface at small (centimeter) spatial scales. Borehole testing data (e.g., hydraulic pumping tests and tracer tests) and indirect observations (e.g., seismic surveys) provide indirect measurements of subsurface properties averaged over much larger (meters to tens of meters) spatial scales. Observations of a given subsurface region using different measurement techniques can yield very different results, and measurements from a single technique can show significant variations over small spatial scales. One of the primary knowledge gaps for model conceptualizations is understanding how to integrate these field observations into the models and parameter estimates. The knowledge gaps include understanding the scale effects and developing methods for data integration that take these effects into account.

Research Needs

Conceptual model development has not been an explicit topic for basic research in its own right. Indeed, conceptual model development is viewed as an inherently empirical and site-specific process using observational approaches that are not easily generalized or tested. The committee believes, however, that basic research that addresses the fundamental approaches and assumptions underlying conceptual model development could produce a tool box of methodologies that are applicable to contaminated sites both inside and outside the DOE complex. This research should focus on the following topics:

1. New observational and experimental approaches and tools for developing conceptual models that apply to complex subsurface environments, including such phenomena as colloidal transport and biologic activity.
2. New approaches for incorporating geological, hydrological, chemical, and biological subsurface heterogeneity into conceptual model formulations at scales that dominate flow and transport behavior.
3. Development of coupled-process models through experimental studies at variable scales and complexities that account for the interacting physical, chemical, and biological processes that govern contaminant fate and transport behavior.
4. Methods to integrate process knowledge from small-scale tests and observations into model formulations, including methods for

incorporating qualitative geological information from surface and near-surface observations into conceptual model formulations.

5. Methods to measure and predict the scale dependency of parameter values.
6. Approaches for establishing bounds on the accuracy of parameters and conceptual model estimates from field and experimental data.

The research needs outlined above call for more hypothesis-driven experimental approaches that address the fundamental methods and assumptions underlying the development of conceptual models. This research will require expertise from a wide range of disciplines and must be conducted at scales ranging from the laboratory bench top to contaminated field sites.

Moreover, to have long-term relevance to the DOE cleanup mission, this research must be focused on the kinds of subsurface environments and contamination problems commonly encountered at major DOE sites. One way to ensure this focus is to give researchers the opportunity to conduct research at contaminated DOE sites. The committee provides additional comments on this issue in the next chapter.[6]

Containment and Stabilization

As noted by DOE in *Paths to Closure* (DOE, 1998a) and as shown in Chapter 2 of this report, a great deal of subsurface contamination is likely to remain at DOE sites even after DOE's cleanup program is completed. It will include contaminant plumes in groundwater, contaminated soil, and waste burial grounds—both the historical burial grounds discussed in Chapter 2 and new burial grounds developed by DOE to dispose of waste from its current and future cleanup operations. DOE is responsible for the long-term management of this contamination and must develop methods to contain and stabilize it until it no longer poses a hazard to humans or the environment—or until new methods to remediate this contamination are developed. DOE's management commitment potentially extends for many thousands of years. DOE's containment and stabilization systems are likely to include surface caps, subsurface barriers, and other in situ stabilization systems. Once installed, these systems will have to be monitored to assure that they perform as expected, and if these systems fail, additional corrective actions may have to be taken to repair the barriers and remediate residual contamination. There has been an increasing emphasis and accep-

[6]See the section titled "Field Sites" in Chapter 6.

tance of waste containment and stabilization in recent years, both in DOE and by regulatory agencies. Decreasing cleanup budgets, evaluations that show that containment is a low-risk choice for some problems, and recognition that some contamination cannot be remediated either with current technologies or conceivable new technologies are responsible for this change in philosophy. This shift in emphasis is perhaps first fully acknowledged by DOE in *Paths to Closure* (DOE, 1998a), which lays out DOE's cleanup objectives, and appears to be a developing trend across the DOE complex.[7] A more recent DOE report (DOE, 1999) discusses the long-term stewardship challenges.

At some sites, containment and stabilization may be an interim measure and has its own set of associated technical problems. These include particularly the availability of appropriate technologies to both contain and stabilize the residual contamination and to monitor and validate the long-term performance of containment and stabilization systems themselves. There is little understanding of the long-term performance of containment and stabilization systems, and there is a general absence of effective methods to validate that such systems are properly installed or that they can provide effective long-term performance. To address this knowledge gap, advances in basic knowledge to support the development of new and improved waste containment and stabilization systems will be needed, as noted below.

The development of improved and novel containment and stabilization approaches will likely have the highest potential for cost savings and lowered risk of the four knowledge gaps identified by the committee. The committee believes that the significant knowledge gaps include the following:

- *Development of robust physical, chemical, and biological containment and stabilization systems.* Traditional containment systems comprised of surface caps, in situ walls, and bottom barriers employ low-permeability materials to reduce water infiltration and provide a barrier to contaminant migration. When designed properly, these systems may provide effective contain-

---

[7]Another recent example of the shift in emphasis to containment strategies can be found in a recent report on disposal of DOE low-level waste (DOE, 1998e). This report shows that DOE's estimates of the volume of its low-level waste requiring disposal between 1998 and 2070 has decreased from about 32 million cubic meters to about 8 million cubic meters, largely because DOE has decided to contain much of this low-level waste in place at its sites, rather than removing it for treatment or disposal elsewhere in the complex. Most of this waste exists in waste burial grounds at the major DOE sites (see Chapter 2).

ment for periods of up to a few decades,[8] but current designs do not meet DOE's needs for containment of its long-lived radioactive and hazardous waste—both for wastes contained in place and new waste sites developed from current and future cleanup operations. Natural low-permeability materials for minimizing infiltration (e.g., clays) work well in humid environments, but they may not be effective in arid regions, where dessication can lead to the development of preferred pathways.

To the committee's knowledge, there has been little or no research or development work on longer-term systems for containment of subsurface contamination of the sort encountered at DOE sites, either by DOE or by other organizations.[9] The knowledge gaps include understanding how to design more effective and permanent barrier systems for long-term containment, especially in arid environments characteristic of the western DOE sites, including the development and application of more durable materials for barrier systems—materials that are compatible with the surrounding environment and with the waste that is being contained.

- *New containment approaches.* Conventional barrier systems seek to minimize water infiltration into the contained waste and to minimize the spread of waste from containment zones into the environment. Surface barrier systems (caps) have proven very effective for retarding water infiltration into containment zones, but they require ongoing maintenance to ensure their continued integrity, and they have short lives relative to the hazard of the contained waste. Moreover, subsurface infiltration barriers (e.g., impermeable walls installed around or beneath waste burial grounds) are extremely difficult to install and maintain, especially barriers emplaced beneath waste containment zones, and their performance is also extremely difficult to monitor.

  New approaches are needed to address DOE's needs for long-term in situ containment and treatment of subsurface contamination. The recent development of reactive barriers (i.e., barriers that degrade or immobilize contaminants through geochemically and biochemically mediated reactions, such as ion

---

[8]To the committee's knowledge, this supposition has never been tested at a DOE site, so the actual longevity of such barrier systems is uncertain at best.

[9]There has been a great deal of research and development work in the United States and other countries on long-term containment systems for spent fuel and vitrified high-level waste, but this work does not appear to be directly applicable to the contamination problems at DOE sites.

exchange or redox processes) is an example of the kind of new approach that holds promise. The continued development of reactive barriers and the development of other hybrid systems (e.g., barrier systems that incorporate biological materials to reduce maintenance requirements and enhance long-term performance, or systems that use controlled water infiltration to enhance waste decomposition or transformation) could improve the technology for containment and in-situ stabilization of subsurface contaminants across the DOE complex.

Research Needs

The construction of stabilization and containment systems is properly within the province of applied technology development and will be the responsibility of other DOE programs (e.g., the Subsurface Contaminants Focus Area). However, basic research focused on the following topics will be needed to support this technology development effort:

1. The mechanisms and kinetics of chemically and biologically mediated reactions that can be applied to new stabilization and containment approaches (e.g., reactions that can extend the use of reactive barriers to a greater range of contaminant types found at DOE sites) or that can be used to understand the long-term reversibility of chemical and biological stabilization methods.
2. The physical, chemical, and biological reactions that occur among contaminants (metals, radionuclides, and organics), soils, and barrier components so that more compatible and durable materials for containment and stabilization systems can be developed.
3. The fluid transport behavior in conventional barrier systems, for example, understanding water infiltration into layered systems, including infiltration under partially saturated conditions and under the influences of capillary, chemical, electrical, and thermal gradients that can be used to support the design of more effective infiltration barrier systems.
4. The development of methods for assessing the long-term durability of containment and stabilization systems.

Monitoring and Validation

The ability to monitor and validate is essential to the successful application of any corrective action to a subsurface contamination problem, as is regulatory acceptance of that action. However, the knowledge and technology bases to support these activities are not fully developed and are receiving little attention in EM's science and technology programs. The monitor process step does not appear on the

Subsurface Contaminants Focus Area's remedial action flow chart (see Figure 3.2), and its validate process step applies only to the confirmation of the performance of a remedial action. As noted in Chapter 3, very little research relevant to these activities is being supported currently by the EM Science Program.

As illustrated by Figure 5.1, monitoring and validation are important at both the front and back ends of the site remediation process. At the front end, monitoring and validation are used to support the development of conceptual and predictive models of subsurface and contaminant behavior (Box 4 and Arrow 1). At the back end, monitoring and validation are used to gain regulatory acceptance for corrective actions and to demonstrate the effectiveness of efforts to remove, treat, or especially to contain contamination (Box 7 and Arrow 2). Such monitoring and validation efforts can also improve the understanding of the contaminant fate and transport processes and can be used to recalibrate and revise conceptual and predictive models—important elements of the model building process.

Improvements in capabilities to monitor and validate could greatly improve the technical success of DOE's efforts to contain and stabilize contamination at its sites. The development of new containment and stabilization approaches could lower the cost, accelerate regulatory approvals for, and increase public confidence in efforts to address DOE contamination problems. In the committee's judgment, the significant knowledge gaps include the following:

- *Design of efficient and effective monitoring systems.* There is little experience with monitoring over the long (decadal to centennial) time scales that are required at DOE sites. Consequently, a great deal of basic knowledge is required to design efficient and effective monitoring systems. The knowledge gaps include understanding what parameters need to be measured to assess system performance (e.g., the performance of a subsurface barrier); where, when, and how to obtain these measurements; and how to relate these measurements to system behavior.
- *Unsaturated zone monitoring.* Monitoring of the unsaturated zone is a special case of the previous knowledge gap and is a special need for DOE, because most of its containment and stabilization systems are being constructed above the water table, especially at the western U.S. sites. Unsaturated zone monitoring is an especially difficult problem; the physics and chemistry of unsaturated zone processes are more complicated than for the saturated zone, and these processes have received far less attention from researchers. Contaminants may be present in both

liquid and gas phases in unsaturated zone environments and under both aerobic and anaerobic conditions. The exchange, degree of equilibration of these phases, and the transport of these phases may occur by different processes with very different rates. There is a great disparity between what can currently be measured and what needs to be measured to predict the behavior of contaminants in many unsaturated settings.

- *Model validation.* A conceptual model is an estimate of the real-world behavior and must be tested to ensure that it appropriately represents the behaviors of interest. This testing is usually carried out by comparing predictions made with the model against field and experimental observations. This testing also allows the model to be improved as new information on the subsurface system is collected. The science of model testing, or validation, has received relatively little attention until recently and is an area where significant work is needed. The knowledge gaps include understanding what measurements need to be collected to validate a model (it is frequently the case that what can be calculated in a model cannot be measured in the field, and vice versa); how to evaluate the relationships between measured and predicted behaviors; and understanding what diagnostic information these differences provide for assessing and improving the accuracy of the models (e.g., see Sidebar 5.2).
- *Performance validation.* Performance validation is a necessary step to document the success, or lack thereof, with every step shown in Figure 5.1. The issues here are similar to those for model validation, that is, how to assess whether the process is performing as designed. The knowledge gaps include understanding what to measure, how to measure it, how to assess discrepancies between designed and measured behavior, and determining what diagnostic information these differences provide for assessing and improving performance.

For example, with regard to locating and characterizing contaminants, one must determine when enough information for risk characterization and remedy selection has been gathered. This requires tools to validate the assessments of contaminant amounts, distributions, and mass release rates. Similar considerations arise for validation of predictive models in the face of variability and uncertainty. The difficulty increases when probability models are introduced to try to deal with uncertainty. With regard to corrective action performance, validation is an essential step that is lacking for many innovative technologies, and has prevented their selection for site remediation because of regulatory and stakeholder concerns. Knowledge gaps in performance validation include understanding how to develop monitoring systems and sampling strategies, understanding the critical system variables that need to be used, strategies for data collection in highly heterogeneous systems, and the development of statistical methods to be used in performance evaluation.

Research Needs

Many of the research needs for monitoring and validation have been covered in previous sections; for example, research on locating contaminants and characterizing contaminant and subsurface properties and research on data integration will provide new knowledge and capabilities for monitoring and validation. In addition, the committee believes that basic research is needed on the following topics:

- *Development of methods for designing monitoring systems to detect both the current conditions and changes in system behaviors.* These methods may involve the application of conceptual, mathematical, and statistical models to determine the types and locations of observation systems and also will involve predicting the spatial and temporal resolutions at which observations need to be made. For example, such methods may help to determine what types of measurements (e.g., core samples from a borehole versus seismic images of the subsurface) can be used to validate the model and also suggest where such measurements should be made in both time and space.
- *Development of validation processes.* The research questions include (1) understanding what a representation of system behavior means and how to judge when a model provides an accurate representation of a system behavior—the model may give the right answers for the wrong reasons and thus may not be a good predictive tool; and (2) how to validate the future performance of the model or system behavior based on present-day measurements. These questions might be addressed through research projects that focus on the development of validation methodologies using real-world examples at DOE sites.
- *Data for model validation.* Determining the key measurements that are required to validate models and system behaviors, the spatial and temporal resolutions at which such measurements must be obtained, and the extent to which surrogate data (e.g., data from lab-scale testing facilities) can be used in validation efforts.
- *Research to support the development of methods to monitor fluid and gaseous fluxes through the unsaturated zone, and for differ-*

entiating diurnal and seasonal changes from longer-term secular changes. These methods may involve both direct (e.g., in situ sensors) and indirect (e.g., using plants and animals) measurements over long time periods, particularly for harsh chemical environments characteristic of some DOE sites. This research should support the development of both the physical instrumentation and measurement techniques. The latter includes measurement strategies and data analysis (including statistical) approaches.

SIDEBAR 5.2 MANAGING UNCERTAINTY

Management of uncertainty in model and performance validation is a theme that cuts across many of the knowledge gaps identified in this chapter. Uncertainties arise in multiple ways. In field data they can emerge in quality features: random measurement error, systematic errors from imperfectly calibrated instruments, and recording and other transmission errors. In mathematical models, uncertainties stem from incorrect specifications and through propagation of errors in the data that are input to the computational models. In the integration (or combination) of models and data, uncertainties are affected by the need to link data and models that are on mismatched scales; some data may have to be aggregated, other data may need to be disaggregated.

Quantifying the uncertainties in, for example, a site characterization problem can involve all the paths described above. There may be several data sets of varying quality; missing data (measurements on some contaminants may be found at some monitoring wells but missing at others); auxiliary data sets (e.g., river flow data) on time scales very different from the frequency of sampled monitoring data; historical records of differing content and quality; and transport models requiring uncertain input parameters. How best to combine the variety of information and assess the accuracy of results and predictions is a great challenge.

Similar issues are found in validation and performance assessment. These may be compounded by the need to perform detailed computer experiments to determine the impacts of uncertainties in data quality and input specifications. For perfomance assessment and validation, attention has to be given to design of future data collection: Where and when should collection be done to assure a desired level of accuracy?

The daunting technical problem is how to respond to complex, though simple sounding, queries (e.g., Where is the contaminant plume now? Where will it be next year?) that demand intricate combinations of computer and statistical models fed by several data sources. Powerful methods such as Bayesian hierarchical modeling are emerging to break such complicated problems into components and, through intensive computation, capture the uncertainties; but implementation is limited by the complexity and scale of the problems typically encountered in subsurface contamination.

Discussion

As noted in the introduction to this chapter, the committee has been selective in the identification of subsurface contamination knowledge gaps and research needs for the EM Science Program. Indeed, the list of knowledge gaps presented in this chapter is not exhaustive and is perhaps notable for what it does not include, namely, the knowledge gaps associated with assessment of risk (Box 5 in Figure 5.1)[10] and many of the corrective actions associated with EM's cleanup program (Boxes 6C through 6E in Figure 5.1). The committee has been selective because (1) it believes that much of the subsurface at DOE sites cannot be remediated cost effectively; (2) the contamination is highly distributed in very large volumes of the subsurface; and (3) the EM Science Program does not have the management or financial capital to support a comprehensive research program to address all of EM's cleanup problems. Further, the committee recognizes that there is much good research on these excluded topics being supported by other programs (see Chapter 4).

The committee has selected the four research areas highlighted in this chapter because, as illustrated by the examples in Chapter 2, these themes cut across all DOE cleanup efforts, and the committee believes that they are key to the long-term success of the DOE's cleanup program. Further, the committee believes that a focused, sustained, and adequately funded research program directed at the knowledge gaps could result both in significant improvements to DOE cleanup capabilities and the effectiveness of its cleanup actions.

The committee discussed whether it should prioritize these four research areas, but decided against doing so. The selection of these four research foci from among a much broader range of potential research areas is in itself a significant prioritization. Further, the committee believes that all four research foci are equally important for DOE's cleanup mission and will need to be pursued aggressively if DOE is to improve its capabilities to address its subsurface contamination problems. The new location, characterization, modeling, and monitoring capabilities that can result from this research, when applied appropriately, will enable DOE to stay on a track that leads to success in its efforts to clean up or contain its widespread subsurface contamination.

---

[10]Of course, the committee recognizes that the basic research needs outlined in this chapter will produce new knowledge on contaminant locations and behavior and thereby affect critical steps in the risk assessment.

6

Recommendations for a Long-Term Research Program

This chapter provides recommendations for a long-term basic research program to address subsurface contamination problems at DOE sites, as directed by the statement of task (see Sidebar 1.1). The recommendations address the following three issues:

1. program vision,
2. research agenda, and
3. implementation of the research agenda.

These recommendations are based on analyses of the information provided in Chapters 2 through 5 of this report and the committee's interim report (NRC, 1998), as well as the reports of the previous National Research Council Committee on the Environmental Management (EM) Science Program (NRC, 1997b).

Program Vision

The EM Science Program has been in existence for almost four years, but there does not appear to be a clear and agreed-upon program vision in DOE, and especially in upper management in the Office of Environmental Management (EM). This conclusion is based on two observations made by the committee during the course of this study. First, the EM Science Program does not appear to be an important part of EM's plan for technology research and development. EM released its Environmental Management Research and Development Plan in 1998 (DOE, 1998f). This plan describes the investments to be made in science and technology to support the DOE cleanup mission. The main text of this plan is 36 pages in length, but only one section comprising two paragraphs is devoted to a discussion of the EM Science Program. The discussion in this plan focuses primarily on the management of the program rather than the program's objectives and content.

Second, the program also does not appear to be a high priority, judging by EM's budget requests to Congress. When the program was created in fiscal year 1996, Congress appropriated $50 million to it from EM's technology development programs.[1] In fiscal year 1997, EM requested $38 million for the program; however, Congress appropriated $50 million, an increase of $12 million over EM's request. In fiscal year 1998, $32 million was requested; Congress again increased its appropriation to $47 million. For fiscal year 1999, $32 million was requested and appropriated.[2] Congress and EM appear to have different views of the importance of this program. In the committee's view, $50 million is inadequate for a research program that has the scope of the EM Science Program. This is especially true since the program was designed to address a wide spectrum of problems, ranging from groundwater contamination to high-level waste. Additional discussions of program funding are provided later in this chapter.

The committee believes that if the program is to remain viable over the long term and have a significant impact on the DOE cleanup mission, program managers must articulate a vision for the program that is supported both programmatically and financially by upper management in EM and DOE. In the committee's view, this vision should include the following four elements:

1. *The objective of the EM Science Program should be to generate new knowledge to support DOE's mission to clean up its contaminated sites.* This objective is consistent with the intent of the congressional language that established this program in 1996 (see Chapter 1), with the conclusions in this committee's interim report (NRC, 1998), and with the conclusions of the previous NRC committee on the EM Science Program (NRC, 1997b). This objective also has been articulated in EM's strategic plan for its science and technology programs (DOE, 1998f). The committee's analysis of subsurface contamination problems in Chapter 2 shows that the environmental remediation and management mission is unlikely to succeed without new knowledge to support the development of new and improved technologies to treat,

---

[1]Specifically, from the Office of Science and Technology.

[2]About $10 million of the appropriation was for research on low-dose ionizing radiation.

remove, or contain and stabilize subsurface contamination at DOE sites.

2. *The EM Science Program should be well connected to DOE's difficult cleanup problems.* In the past, the program was operated somewhat like a first generation research and development program, which has been characterized (perhaps with tongue in cheek) as "put a few bright people in a dark room, pour in money, and hope" (Hamel and Prahalad, 1989; see also Roussel and others, 1991). Clearly, this model is inappropriate for the program, which will succeed only if it is well connected, both in perception and fact, to EM's significant cleanup problems. The efforts by the program managers to develop science plans represents a positive move in this direction. The committee hopes the recommendations for basic research foci presented in this report will aid this effort.
3. *A major focus of the EM Science Program should continue to be on research to resolve DOE's subsurface contamination problems.* Based on its review of subsurface contamination problems in Chapter 2, the committee concluded that DOE faces significant difficulties in remediating radionuclide-, metal-, and solvent-contaminated soil and groundwater at all of its major sites. DOE's own analyses and publications (see Chapter 2) also support the conclusion that subsurface contamination is a significant long-term problem. Moreover, previous National Research Council reports have shown that DOE lacks the technologies needed to effectively remediate much of the subsurface contamination at its sites (e.g., NRC, 1997a,c, 1999). The committee believes, therefore, that new knowledge (and technologies) will be required to address DOE's subsurface contamination problems, and the committee recommends that subsurface contamination should continue to be a major focus of the EM Science Program.
4. *The EM Science Program should have a long-term, multidisciplinary, basic-research[3] focus.* As discussed in Chapter 2, the active phase of DOE's cleanup efforts is planned to last until at least 2050 (see Table 2.3), and DOE faces additional long-term monitoring commitments beyond 2070. Consequently, DOE has sufficient time to undertake and to benefit from long-term basic research under the auspices of the EM Science Program.

---

[3]As noted in Sidebar 1.1, basic research creates new generic knowledge and is focused on long-term, rather than short-term, problems.

A long-term basic research focus will allow the EM Science Program to sponsor fundamental research on subsurface contamination that can lead to significant knowledge and technology breakthroughs.[4] It also will insulate the program from the ongoing shifts in emphasis in the DOE cleanup effort. Indeed, a long-term focus will enable the program to provide sustained funding, including renewals in funding for successful projects, so that investigators can pursue and build on significant research results.

The committee defines "long term" as long enough to set ambitious goals for addressing the knowledge gaps identified in Chapter 5 and to have reasonable expectations that those goals can be attained. In the committee's judgment, a time horizon on the order of a decade will be required to make cumulative progress on the knowledge gaps, although shorter-term results of use to DOE almost certainly will be obtained over the lifetimes of individual research projects (i.e., over a three-year time frame). A decadal time horizon would produce a critical mass of researchers and research projects focused on the knowledge gaps; it would provide for several proposal cycles so that investigators could pursue important research ideas and develop significant research results. With the proper encouragement from program managers, it would also encourage researchers to develop collaborations that could lead to novel approaches to addressing the knowledge gaps, many of which are highly interdisciplinary.[5]

The decadal time horizon would allow investigators to apply for competitive renewals to pursue significant research findings. Such renewals could accelerate progress in addressing the knowledge gaps, keep good researchers focused on problems of importance to DOE, and, in the case of university-funded projects, provide a strategic investment in future generations of researchers knowledgeable of DOE's problems through support for graduate students and postdoctoral scientists.

Research Agenda

The committee has identified four critical knowledge gaps that it believes are significant impediments to the successful completion of DOE's cleanup mission and that are addressable through a focused, sustained, and adequately funded research program. Although these

---

[4]A good discussion of the value of long-term research is provided in a previous NRC report (NRC, 1997b), especially in Chapter 2.

[5]The value of multidiscipinary research is discussed in some detail in a previous NRC report (NRC, 1997b).

may not be the only critical knowledge gaps in DOE's evolving subsurface cleanup program, the committee is certain that these four challenges must be addressed for DOE's cleanup program to be completed safely and cost effectively.

The committee recommends that the subsurface component of the EM Science Program should emphasize research on the four knowledge gaps that were identified in Chapter 5 and discussed below. The specific research topics suggested in this section are for illustrative purposes and are not meant to be prescriptive. The committee expects that the research supported by this program will be truly basic, imaginative, and innovative. The committee's recommendation of four research foci does not imply that the subsurface research supported in the current EM Science Program portfolio is inappropriate or misdirected. Rather, these four foci represent areas where more research clearly is needed.

Location and Characterization of Subsurface Contaminants and Characterization of the Subsurface

The challenges of locating subsurface contamination are magnified by the wide range of contaminant types (e.g., mixtures of organic solvents, metals, and radionuclides) in the subsurface at many DOE sites; the wide variety of geological and hydrological conditions across the DOE complex; and the wide range of spatial resolutions at which this contamination must be located and characterized, from widely dispersed contamination in groundwater plumes to small isolated hot spots in waste burial grounds.

As discussed in Chapter 5, the committee believes that basic research is needed to support the development of the following capabilities to locate and characterize contamination in the subsurface and to characterize subsurface properties at the scales that control contaminant fate and transport behavior:

- Improved capabilities for characterizing the physical, chemical, and biological properties of the subsurface.
- Improved capabilities for characterizing physical, chemical, and biological heterogeneity, especially at the scales that control contaminant fate and transport behavior. Approaches that allow the identification and measurement of the heterogeneity features that control contaminant fate and transport to be obtained directly (i.e., without having to perform a detailed characterization of the subsurface) are especially needed.
- Improved capabilities for measuring contaminant migration and system properties that control contaminant movement.
- Methods to integrate data collected at different spatial and tem-

poral scales to better estimate contaminant and subsurface properties and processes.

- Methods to integrate such data into conceptual models.

Conceptual Modeling

Existing conceptual and predictive models have often proven ineffective for understanding and predicting contaminant movement, especially at sites that have thick unsaturated zones or complex subsurface characteristics. Accurate conceptualizations are essential for understanding the long-term fate of contaminants in the subsurface and the selection and application of appropriate corrective actions. The committee believes that basic research explicitly focused on fundamental approaches and assumptions underlying conceptual model development could produce a tool box of methodologies that are applicable to contaminated sites both inside and outside the DOE complex. This research should focus on the following topics:

- New observational and experimental approaches and tools for developing conceptual models that apply to complex subsurface environments, including such phenomena as colloidal transport and biologic activity.
- New approaches for incorporating geological, hydrological, chemical, and biological subsurface heterogeneity into conceptual model formulations at scales that dominate flow and transport behavior.
- Development of coupled-process models through experimental studies at variable scales and complexities that account for the interacting physical, chemical, and biological processes that govern contaminant fate and transport behavior.
- Methods to integrate process knowledge from small-scale tests and observations into model formulations, including methods for incorporating qualitative geological information from surface and near-surface observations into conceptual model formulations.
- Methods to measure and predict the scale dependency of parameter values.
- Approaches for establishing bounds on the accuracy of parameters and conceptual model estimates from field and experimental data.

The research needs outlined above call for more hypothesis-driven experimental approaches that address how to integrate the understanding of system behavior. This research will require expertise from a wide range of disciplines and must be conducted at scales ranging from the laboratory bench top to contaminated field sites. Moreover, to have long-term relevance to the DOE cleanup mission, this research must be focused on the kinds of subsurface environments and contamination problems commonly encountered at major DOE sites.

Containment and Stabilization

There has been an increasing emphasis on and acceptance of waste containment and stabilization in recent years, both in DOE and by regulatory agencies. Decreasing cleanup budgets, evaluations that show containment is a low-risk choice for some problems, and recognition that some contamination cannot be remediated either with current technologies or conceivable new technologies are responsible for this change in philosophy. However, at some sites, containment and stabilization may be an interim measure and has its own set of associated technical problems. There is little understanding of the long-term performance of containment and stabilization systems, and there is a general absence of robust and cost-effective methods to validate that such systems are installed properly or that they can provide effective long-term protection.

The construction of stabilization and containment systems is properly within the province of applied technology development. However, basic research focused on the following topics will be needed to support this technology development effort:

- The mechanisms and kinetics of chemically and biologically mediated reactions that can be applied to new stabilization and containment approaches (e.g., reactions that can extend the use of reactive barriers to a greater range of contaminant types found at DOE sites) or that can be used to understand the long-term reversibility of chemical and biological stabilization methods.
- The physical, chemical, and biological reactions that occur among contaminants (metals, radionuclides, and organics), soils, and barrier components so that more compatible and durable materials for containment and stabilization systems can be developed.
- The fluid transport behavior in conventional barrier systems, for example, understanding water infiltration into layered systems, including infiltration under partially saturated conditions and under the influences of capillary, chemical, electrical, and thermal gradients can be used to support the design of more effective infiltration barrier systems.
- The development of methods for assessing the long-term durability of containment and stabilization systems.

Monitoring and Validation

Monitoring and validation are necessary at both the front and the back ends of the site remediation process. At the front end, monitoring and validation are used to support the development of conceptual and predictive models of subsurface and contaminant behavior. At the back end, monitoring and validation are used to demonstrate the effectiveness of efforts to remove, treat, or especially to contain contamination and to gain regulatory acceptance for such corrective actions. Moreover, such monitoring and validation efforts can also improve the understanding of the contaminant fate and transport processes and can be used to recalibrate and revise conceptual and predictive models—important elements of the model building process.

The ability to monitor and validate is essential to the successful application of any corrective action to a subsurface contamination problem and regulatory acceptance of that action. However, the knowledge and technology bases to support these activities are not fully developed and are receiving little attention in EM's science and technology programs.

Many of the research opportunities for monitoring and validation have been covered in the research emphases discussed above. In addition, the committee believes that basic research is needed on the following topics:

- Development of methods for designing monitoring systems to detect both current conditions and changes in system behaviors. These methods may involve the application of conceptual, mathematical, and statistical models to determine the types and locations of observation systems and prediction of the spatial and temporal resolutions at which observations need to be made.
- Development of validation processes. The research questions include (1) understanding what a representation of system behavior means and how to judge when a model provides an accurate representation of a system behavior—the model may give the right answers for the wrong reasons and thus may not be a good predictive tool; and (2) how to validate the future performance of the model or system behavior based on present-day measurements.
- Data for model validation. Determining the key measurements that are required to validate models and system behaviors, the spatial and temporal resolutions at which such measurements must be obtained, and the extent to which surrogate data (e.g.,

data from lab-scale testing facilities) can be used in validation efforts.

- Research to support the development of methods to monitor fluid and gaseous fluxes through the unsaturated zone, and for differentiating diurnal and seasonal changes from longer-term secular changes. These methods may involve both direct (e.g., in situ sensors) and indirect (e.g., using plants and animals) measurements over long time periods, particularly for harsh chemical environments characteristic of some DOE sites. This research should support the development of both the physical instrumentation and measurement techniques. The latter includes measurement strategies and data analysis (including statistical) approaches.

Other Recommendations on the Research Agenda

Within the four research emphases described above, the committee recommends that the EM Science Program encourage research on metals and radionuclides. Many of the metal and radionuclide contamination problems are almost wholly "owned" by DOE, especially transuranic contaminants. The committee recognizes that DOE also has many dense non-aqueous phase liquid contamination problems at its sites, but as discussed in Chapter 4, there are many research programs in other parts of DOE and in other federal agencies that provide funding for research on this contaminant. The committee judges that this is less true for research on metals and radionuclides.

The committee also recommends that there be sufficient flexibility in future calls for subsurface proposals so that support can be provided for high-risk but potentially high-payoff research ideas that intersect with the four research emphases. Such projects could produce major knowledge breakthroughs leading to significant improvements in DOE's cleanup capabilities and costs.

Implementation of the Research Agenda

The EM Science Program is a basic research program focused on very real DOE problems. The program's success will be measured both by its impact on advancing the science and its impact on DOE site cleanup. To be successful, the program must not only be focused on the right problems but it must also encourage researchers to do the right work; and it must find a way to hand off the results of this work to technology developers and problem holders at DOE sites. In this section the committee offers strategic recommendations for achieving the first two objectives;[6] these recommendations address the following:

- integration,
- field sites, and
- program funding.

Integration

The committee believes that EM Science Program managers must encourage and support integration activities across the program if it is to advance subsurface science and have a significant impact on DOE cleanup. To this end, the program's implementation strategy should have the following three integrative elements:

1. *The program should continue to reach beyond the usual group of DOE researchers to pull in new and novel ideas to address DOE-specific problems.* Much of the expertise needed to address the knowledge gaps identified in Chapter 5 can be found outside the traditional DOE research community. Indeed, the previous NRC committee on the EM Science Program encouraged program managers to broaden the community of investigators involved in the program and to expand the core or committed cadre of investigators who are knowledgeable of EM's problems (NRC, 1997b, p. 4). Judging from the committee's review of the current program portfolio in Chapter 3, the program appears to be making progress in meeting this objective. The committee encourages program managers to continue their efforts to broaden the community of researchers from government agencies with research capabilities, national laboratories, universities, and industry.
2. *The program should continue to encourage multidisciplinary research and university-national laboratory-industry collaborations that will promote new insights into the very complex subsurface problems at DOE sites.* Many of the challenges identified in Chapter 5 are technically difficult and inherently interdisciplinary. The committee believes that to make significant progress in addressing them, the program must encourage and support multidisciplinary research teams. There is a good representation of multiple-investigator projects in the current program portfolio (see Table 3.1), especially collaborations among university and

[6]The third objective on moving science into application, although extremely important, is beyond the statement of task for the present study.

national laboratory scientists. The committee recommends that program managers continue to encourage such collaborations by providing support for workshops and seminars to bring scientists together with site problem holders to discuss DOE contamination problems and possible research approaches; this was recommended by the previous NRC committee on the EM Science Program (NRC, 1997b). The committee offers an additional recommendation to encourage collaborations in the next section of this chapter.

3. *The program should integrate existing data and ideas—both from DOE sites and basic research programs outside DOE—to promote advancements in subsurface science and improvements in capabilities to address DOE's subsurface contamination problems.* The program can also play a lead role in integrating the considerable amount of relevant subsurface science research that is being supported by DOE and other federal agencies. As discussed in Chapter 4, there is a great deal of potentially relevant subsurface research that is being supported outside the EM Science Program, but the committee found that there is little or no effort being made to coordinate these research investments or to transfer results into the DOE cleanup program.

   The program has the potential to provide leadership in the advancement of subsurface science, primarily because it can provide access to scientifically interesting and intellectually challenging problems at DOE sites—problems that do not exist anywhere else in the United States and few places in the world—and because many DOE sites possess rich caches of data that can be used in research projects to address the knowledge gaps identified earlier in this chapter. Groundwater monitoring data from sites like Hanford, for example, could be used to develop forensic methods to estimate contaminant release rates or to develop and test conceptual models (see Chapter 2). However, to be useful in this regard, researchers must have access to DOE data, sites, and site-knowledgeable personnel.

Field Sites

The committee recommends that program managers examine the feasibility of developing field research sites where investigators with program awards could work on the knowledge gaps described earlier in this chapter. These field sites could include contaminated or uncontaminated areas at the major DOE sites; analog uncontaminated sites that have subsurface characteristics similar to contaminated DOE sites; and even virtual sites comprised of data on historical and contemporary contamination problems. These sites could be established by the program itself or in cooperation with other research programs.[7]

Access to field research sites could allow investigators to make significant progress in addressing the four knowledge gaps identified previously in this chapter. For example, research on location and characterization will require access to field sites where measurements on real subsurface and contaminant properties can be made and where measurement methodologies can be compared. Research on conceptual model development and testing and on validation and monitoring are inherently field based. Researchers must have hands-on familiarity that comes from working in the field to develop and test new methodologies and approaches. Research on containment and stabilization will require access to field sites to test ideas developed in the laboratory or modeling studies, for example, to measure in situ rates of chemical reactions that could be used to develop new and improved containment and treatment approaches.

The establishment of field research sites could have several tangible benefits to the program. First, program managers could encourage research on specific knowledge gaps by establishing field sites in certain kinds of contaminated environments. For example, program managers could encourage research on unsaturated zone contamination by establishing an unsaturated zone field site at one of the major DOE installations in the western United States. Second, such sites could attract new researchers to the program, especially if the field sites could provide research opportunities unavailable through other programs. Third, field sites could encourage both formal and informal multidisciplinary collaborations among the researchers working at these sites, thereby providing benefits that are greater than the sum of individual projects. Such collaborations could be enhanced if the program identified a site manager who could coordinate the research activities at the site and encourage researchers with common interests to work together. Finally, the establishment of field research sites could facilitate the transfer of research results into application because of site proximity to the problem holders and the problems themselves.

The establishment of field research sites is potentially expensive, especially if the field sites are located in contaminated areas where drilling, sample collection, and sample handling would be costly and where investigators would be required to follow DOE environmental

---

[7]For example, the Natural and Accelerated Bioremediation Research program (see Chapter 4) plans to establish a field research site at a major DOE site in fiscal year 2000 and may be an appropriate test bed for research sponsored by the EM Science Program.

health and safety procedures.[8] Moreover, the program may have to pay for the services of a site manager and may have to develop procedures and provide funding to ensure that site data are properly archived and disseminated to researchers and cleanup personnel. Consequently, the establishment of such sites would require additional budget support beyond that required to fund individual research projects, and well beyond the amount of funding available to the program for new starts in fiscal year 1999. Indeed, support for field research sites could consume a significant fraction of the program budget for new starts. However, field research is just one component in a well-balanced research program and should not be supported at the expense of projects that involve laboratory and modeling approaches. Consequently, additional funding would have to be made available to the program to support the development of field sites, or funding for the sites would have to come from other parts of EM (e.g., the Office of Site Closure or the Office of Project Completion, which have the primary responsibility for cleanup of contaminated soil and groundwater). The use of such sites would have to be evaluated periodically to determine whether they are adding value to the research effort, particularly given the cost of such sites relative to the total size of the program budget.

Program Funding

The issue of funding for the EM Science Program has received a great deal of attention from a previous NRC committee (NRC, 1997b), which concluded that the "program must be large enough to support a significant number of 'new starts' (i.e., new projects or competitive renewals) each year if it is to be successful in attracting innovative proposals from outstanding researchers ...." The program needs to have a significant number of new starts each year to keep potential investigators engaged and willing to invest the time and intellectual energy to become knowledgeable of DOE problems and develop research ideas to address them.

New starts will help establish a cadre of knowledgeable and committed investigators—undergraduates, graduates, postdocs, and professionals—who can be called on by DOE in the years ahead for help with its most difficult contamination problems. New starts are also needed to maintain continuity in the research effort; the advancement of scientific knowledge is a cumulative effort involving many scientists

[8]Field research at a contaminated site would need to be carefully reviewed by managers familiar with the research activity and the nature of hazards at the site to ensure that health and safety requirements are met and that the research activity does not exacerbate the spread of contamination.

over long periods of time. This effort is set back significantly each time program funding is interrupted. Researchers may become frustrated and move on to other projects, and graduate students and postdocs may seek training in other fields. Even a single year's interruption in program support can have negative effects that last for several years.

Small program budgets can also lead to significant investigator frustration, especially when proposal success rates fall below accepted norms and highly rated proposals are declined. When proposal success rates fall to low levels, talented investigators may view the proposal preparation and submission process as a bad investment of their time and may stop submitting proposals. This will have an immediate negative impact on the quality of the research being sponsored and long-term negative impact on the DOE technology development efforts.

It is the committee's strong impression that the current level of program funding is not sufficient to support the research emphases outlined in this report, especially when subsurface research is just one of many research areas supported by the program. However, the committee has no basis on which to recommend a specific funding level, and such a recommendation would be well beyond the committee's statement of task. The committee believes that it is the responsibility of EM Science Program managers to estimate the amount of funding required to provide adequate support for a research program focused on addressing the knowledge gaps presented in Chapter 5. One approach for estimating the annual research budget is to estimate the number of projects needed to attain a critical mass of research on each technical challenge area, and then to multiply that number by the average annual grant size.

The committee believes that such estimates could be used to justify future, and possibly larger, budget requests to upper DOE management and Congress, especially if the estimates were reviewed and validated by DOE's internal advisory committees like the Environmental Management Advisory Board or other external advisory committees. Future and larger budget requests are likely to be seen in an increasingly more favorable light as the EM Science Program becomes more firmly connected to EM's cleanup problems.

Concluding Observations

The basic research supported by the EM Science Program and other relevant research programs in the federal government will have little if any impact on DOE cleanup unless research results are transferred into technology development programs in EM and to problem holders at

DOE sites. EM Science Program managers have a responsibility to ensure that specific procedures are in place to foster the handoff from research to development, both for research results developed in its programs and from other relevant programs in the federal government.

The committee believes that there must be strong scientific, technical, and management leadership at all levels of EM, from the EM Science Program up to and including the assistant secretary for environmental management, if significant progress on closing the knowledge gaps and applying results effectively to the cleanup effort is to be made in the next decade. The development of such leadership remains a continuing challenge—and a significant opportunity—for the EM Science Program and DOE.

References

Agnew, S.F., J. Boyer, R.A. Corbin, T.B. Duran, J.R. Fitzpatrick, K.A. Jurgensen, T.P. Ortiz, and B.L. Young. 1997. Hanford Tank Chemical and Radionuclide Inventories: HDW Model, Rev. 4. Report LA-UR-96-3860. Los Alamos National Laboratory, New Mexico.

Albright, J.N., and D.S. Dreesen. 2000. Microhole technology lowers reservoir exploration, characterization costs. Oil and Gas Journal (January 10):39-41.

DOE (Department of Energy). 1995. Closing the Circle on the Splitting of the Atom: The Environmental Legacy of Nuclear Weapons Production in the United States and What the Department of Energy Is Doing About It. DOE/EM-0266. Washington, D.C.: Office of Environmental Management.

DOE. 1996. The 1996 Baseline Environmental Management Report (2 volumes). DOE/EM-0290. Washington, D.C.: Office of Environmental Management.

DOE. 1997a. Linking Legacies: Connecting the Cold War Nuclear Weapons Production Processes to Their Environmental Consequences. DOE/EM-0319. Washington, D.C.: Office of Environmental Management.

DOE. 1997b. Subsurface Contaminants Focus Area 1997 Annual Report. DOE/EM-0361. Washington, D.C.: Office of Environmental Management.

DOE. 1998a. Accelerating Cleanup: Paths to Closure. DOE/EM-0362. Washington, D.C.: Office of Environmental Management.

DOE. 1998b. Groundwater/Vadose Zone Integration Project Specification. DOE/RL-98-48. Draft C. December 17. Richland, Washington: Department of Energy.

DOE. 1998c. Environmental Management Science Program Workshop. CONF-980736. Washington, D.C.: Office of Environmental Management.

DOE. 1998d. Subsurface Contaminants Focus Area 1998 Annual Report. Washington, D.C.: Office of Environmental Management.

DOE. 1998e. The Current and Planned Low-Level Waste Disposal Capacity Report, Revision 1. Washington, D.C.: Office of Environmental Management.

DOE. 1998f. Environmental Management Research and Development Plan: Solution-Based Investments in Science and Technology. Washington, D.C.: Office of Environmental Management.

DOE. 1998g. Report to Congress on the U.S. Department of Energy's Environmental Management Science Program: Research Funded and Its Linkages to Environmental Cleanup Problems (3 volumes). DOE/EM-0357. Washington, D.C.: Office of Environmental Management.

DOE. 1999. From Cleanup to Stewardship. DOE/EM-0466. Washington, D.C.: Office of Environmental Management.

EPA (Environmental Protection Agency). 1977. Cleaning Up the Nation's Waste Sites: Markets and Technology Trends. EPA 542-R-96-005. Washington, D.C.: Office of Solid Waste and Emergency Response.

Gephart, R.E., and R.E. Lundgren. 1998. Hanford Tank Clean Up: A Guide to Understanding the Technical Issues. Richland, Washington: Battelle Press.

Hamel, G., and C.K. Prahalad. 1989. Strategy and Intent. Harvard Business Review (May-June): 63.

Hunt, J.R., N. Sitar, and K.S. Udell. 1986a. Nonaqueous phase liquid transport and cleanup: I. analysis of mechanisms. Water Resources Research 24(8): 1247-1258.

Hunt, J.R., N. Sitar, and K.S. Udell. 1986b. Nonaqueous phase liquid transport and cleanup: II. experimental studies. Water Resources Research 24(8): 1259-1269.

Illangasekare, T.H., J.L. Ramsey, K.S. Jensen, and M.B. Butts. 1995. Experimental study of movement and distribution of dense organic contaminants in heterogenous aquifers. Journal of Contaminant Hydrology 20: 1-25.

INEEL (Idaho National Engineering and Environmental Laboratory). 1997. Decision Analysis for Remediation Technologies (DART) Data Base and User's Manual. INEEL/EXT-97-01052, Idaho Falls, Idaho: INEEL.

Karsting, A. B., D.W. Efurd, D.L. Finnegan, D.J. Rokop, D.K. Smith, and J.L. Thompson. 1999. Migration of plutonium in ground water at the Nevada Test Site. Nature 397: 56-59.

Kueper, B.H., and E.O. Frind. 1991. Two-phase flow in heterogeneous porous media 1. model development. Water Resources Research 27(6): 1049-1057.

NRC (National Research Council). 1966. Committee on Geological Aspects of Radioactive Waste Disposal: Report to the U.S. Atomic Energy Commission. Washington, D.C.: National Academy Press.

NRC. 1994. Alternatives for Ground Water Cleanup. Washington, D.C.: National Academy Press.

NRC. 1995. Allocating Federal Funds for Science and Technology. Washington, D.C.: National Academy Press.

NRC. 1996a. Building an Effective Environmental Management Science Program: Initial Assessment. Washington, D.C.: National Academy Press.

NRC. 1996b. Letter Report on the Environmental Management Science Program Dated October 6. Washington, D.C.: Board on Radioactive Waste Management, National Research Council.

NRC. 1997a. Innovations in Ground Water and Soil Cleanup: From Concept to Commercialization. Washington, D.C.: National Academy Press.

NRC. 1997b. Building an Effective Environmental Management Science Program: Final Assessment. Washington, D.C.: National Academy Press.

NRC. 1997c. Improving the Environment: An Evaluation of DOE's Environmental Management Program. Washington, D.C.: National Academy Press.

NRC. 1998. Letter Report on the Environmental Management Science Program Dated December 10. Washington, D.C.: Board on Radioactive Waste Management, National Research Council.

NRC. 1999. Ground Water and Soil Cleanup: Improving Management of Persistent Contaminants. Washington, D.C.: National Academy Press.

Olsen, C.R., I.L. Larsen, P.D. Lowry, C.R. Moriones, C.J. Ford, K.C. Dearstone, R.R. Turner, B.L. Kimmel, and C.C. Brandt. 1992. Transport and accumulation of cesium-137 and mercury in the Clinch River and Watts Bar Reservoir System. ORNL/ER-7. Oak Ridge, Tennessee: Oak Ridge National Laboratory.

Pfannkuch, H.O. 1984. Determination of the contaminant source strength from mass exchange processes at the petroleum ground-water interface in shallow aquifer systems. In Proceedings of the NWWA Conference on Petroleum Hydrocarbons and Organic Chemicals in Groundwater, pp. 111-129. Dublin, Ohio: National Well Water Association.

Pielke, R.A., and R. Byerly. 1998. Beyond basic and applied. Physics Today (February): 42-46.

Reidel, S.P., K.A. Lindsey, and K.R. Fecht. 1992. Field Trip Guide to the Hanford Site. WHC-MR-0391. Richland, Washington: Westinghouse Hanford Co.

Riley R.G., and J.M. Zachara. 1992. Chemical Contaminants on DOE Lands and Selection of Contaminated Mixtures for Subsurface Science Research. DOE/ER-0547T. Washington, D.C.: Office of Energy Research.

Roussel, P.A., K.N. Saad, and T.J. Erickson. 1991. Third Generation R&D: Managing the Link to Corporate Strategy. Boston, Massachusetts: Harvard Business School Press.

Sandia National Laboratories. 1996. Performance Evaluation of the Technical Capabilities of DOE Sites for Disposal of Mixed Low-Level Waste. Albuquerque, New Mexico: Sandia National Laboratories.

Schwille, F. 1988. Dense Chlorinated Solvents in Porous and Fractured Media. (Translated by J. F. Pankow). Chelsea, Michigan: Lewis Publishers.

Thorpe, R.K., W.F. Isherwood, M.D. Dresen, and C.P. Webster-Sholten (eds.). 1990. CERCLA Remedial Investigations Report for the LLNL Livermore Site. UCAR-10299. Livermore, California: Lawrence Livermore National Laboratory.

Tompson, A.F.B., R.D. Falgout, S.G. Smith, W.J. Bosl, and S.F. Ashby, 1998. Analysis of subsurface contaminant migration and remediation using high performance computing. Advances in Water Resources 23(3): 203-221. (This paper can be viewed in electronic form at http://www.elsevier.nl/locate/advwatres, where a number of animated sequences are available to more clearly visualize the effects of heterogeneity.)

Walker, J.B., and P.J. Liebendorfer. 1998. Long-term stewardship at the Nevada Test Site. Paper prepared for the State Tribal Government Working Group Subcommittee on Stewardship. (http://207.12.87.1/nucwaste/nts/steward.htm#contamination)

Westinghouse Savannah River Company. 1995. A/M Area Groundwater Cleanup. Fact Sheet. Aiken, South Carolina: Savannah River Operations Office.

Westinghouse Savannah River Company. 1998. Savannah River Site Burial Ground Complex. Fact Sheet. Aiken, South Carolina: Savannah River Operations Office.

Appendixes

A

Description of the Environmental Management Science Program

The Environmental Management (EM) Science Program was initiated by the 104th Congress to stimulate basic research and technology development for cleanup of the DOE complex. The program was created in the conference report that accompanied the Energy and Water Development Appropriations Bill (Public Law 104-46, 1995):

> *The conferees agree with the concern expressed by the Senate that the Department [of Energy] is not providing sufficient attention and resources to longer term basic science research which needs to be done to ultimately reduce cleanup costs. The current technology development program continues to favor near-term applied research efforts while failing to utilize the existing basic research infrastructure within the Department and the Office of Energy Research. As a result of this, the conferees direct that at least $50,000,000 of the technology development funding provided to the environmental management program in fiscal year 1996 be managed by the Office of Energy Research and used to develop a program that takes advantage of laboratory and university expertise. This funding is to be used to stimulate the required basic research, development and demonstration efforts to seek new and innovative cleanup methods to replace current conventional approaches which are often costly and ineffective.*

The EM Science Program is managed jointly by DOE's Office of Environmental Management and Office of Science.[1] Staff from these two offices work together to develop proposal calls, review proposals,

---

[1] Formerly the Office of Energy Research. The office was renamed by Congress in 1998.

and make award recommendations. Staff of these two offices have different but complementary roles in the proposal solicitation and review process, as explained below.

The program is run on an annual cycle that begins each fall with the publication of a program announcement in the Federal Register inviting investigators in academia, national laboratories, and industry to submit proposals to the program. The proposal submission process has two steps. Initially, investigators are invited to submit short descriptions of their research ideas, or pre-proposals, for consideration.[2] These pre-proposals undergo an in-house screening to determine whether they meet the criteria laid out in the program announcement, namely, whether the proposed project constitutes basic research (as opposed to technology development, for example) and addresses one or more of the identified priority areas. Investigators whose pre-proposals are judged to meet these criteria are then encouraged to submit full proposals.

The review of full proposals is carried out in a two-stage process, the first to assess scientific merit and the second to assess program relevance. This review process is managed jointly by Office of Science and Office of Environmental Management staff. Merit review is obtained through peer review panels, composed of scientists from industry, national laboratories, and universities, organized along disciplinary lines consistent with normal Office of Science practices. Those proposals that are highly rated in the merit review are then put forward for relevance review, which is performed by a panel of program managers from DOE head-quarters and field offices who are knowledgeable of EM's cleanup needs and priorities.

Following these reviews, Office of Science and Office of Environmental Management program staff provide an overall rating for each of the proposals and make award recommendations to their management. Final award decisions are made by the director of the Office of Science and the deputy assistant secretary for science and technology, Office of Environmental Management. Successful proposals are funded for up to three years, typically at $100,000 to $300,000 per year.

[2]The preapplication process is voluntary.

B

List of Presentations

September 9, 1998

Opening Remarks by DOE, Gerald Boyd, DOE-Office of Environmental Management (EM); Roland Hirsch, DOE-Office of Energy Research (ER)

Overview of the EM Science Program, Mark Gilbertson, DOE-EM; Roland Hirsch, DOE-ER

Overview of the DOE Complex and Subsurface Contamination Problems, Tom Hicks, Savannah River Site

Overview of the EM Science Program Portfolio Directed at Subsurface Contamination Problems, Tom Williams, Idaho Engineering and Environmental Laboratory

Other R&D Work in EM Focused on Subsurface Contamination Problems, Tom Hicks, Savannah River Site

November 10, 1998

Update on EM Science Program Budget, Mark Gilbertson, DOE-EM

Subsurface Contamination Problems at the Savannah River Site, Tom Temples, DOE-Savannah River

Subsurface Contamination Problems at the Oak Ridge Site, Gary Hartman and Paula Kirk, Oak Ridge Site

December 15, 1998

Subsurface Contamination Problems at the Nevada Test Site, Robert Bangerter, DOE-Nevada Operations Office

Subsurface Contamination Problems at the Idaho Site, Tom Williams, Tom Wood, Tom Stoops, Bob Smith, Annette Schafer, Idaho Engineering and Environmental Laboratory

Subsurface Contamination Problems at the Hanford Site, Roy Gephart, John Zachara, Pacific Northwest National Laboratory

January 28, 1999

Update on EM Science Program, Gerald Boyd, DOE-EM; Mark Gilbertson, DOE-EM; Roland Hirsch, DOE-Office of Science (SC)

EM Science Program Opportunities and Challenges in Subsurface Research: A View from Environmental Management Advisory Board (EMAB), Frank Parker, Vanderbilt University, EMAB Science Committee Chair

Research Programs in the U.S. Geological Survey (USGS), Mary Jo Baedecker, USGS

Research Programs in the U.S. Environmental Protection Agency (EPA), Lee Mulkey, EPA

Research Programs in the U.S. Department of Energy, Skip Chamberlain, DOE-EM and John Houghton, DOE-SC

Research Programs in the U.S. Department of Defense, Bradley Smith, Strategic Environmental Research and Development Program

May 6, 1999

Update on the EM Science Program and Desired Outcomes of the Committee's Work, Mark Gilbertson, DOE-EM

C

Biographical Sketches of Committee Members

JANE C.S. LONG (Chair) is dean of the Mackay School of Mines at the University of Nevada, Reno. She is an expert in fracture hydrology and has worked on several U.S. and international underground nuclear repository research projects. She serves on the National Research Council's Board on Radioactive Waste Management and has served as chair of the Board on Earth Science's Rock Mechanics Committee. Dr. Long received an Sc.B. in engineering from Brown University, an M.S. in geotechnical engineering and a Ph.D. in materials science and mineral engineering from the University of California, Berkeley.

JAMES K. MITCHELL (Vice-Chair) is university distinguished professor emeritus at Virginia Polytechnic Institute and State University. He has served on several NRC committees including the Committee on Seeing Into the Earth and as chair of the Geotechnical Board. Dr. Mitchell's expertise lies in the areas of soil behavior related to geotechnical problems, soil improvement and ground reinforcement, and in situ measurement of soil properties. He received his B.S. in civil engineering from Rensselaer Polytechnic Institute, and his M.S. and Sc.D. in civil engineering from Massachusetts Institute of Technology. He is a member of the National Academy of Sciences and the National Academy of Engineering.

RANDALL J. CHARBENEAU is professor of civil engineering and associate dean for research in the College of Engineering at the University of Texas at Austin. His expertise is in groundwater pollution, fate and transport, and modeling. Dr. Charbeneau is a member of the NRC Committee on Technologies for Cleanup of Subsurface Contaminants in the DOE Weapons Complex. He holds civil engineering degrees from the University of Michigan (B.S.), Oregon State University (M.S.), and Stanford University (Ph.D.).

JEFFREY J. DANIELS is an associate professor in the Department of Geological Sciences at Ohio State University. His expertise is in shallow geophysics for subsurface characterization, and he focuses his research on the use of ground penetrating radar and shallow seismic techniques for remote characterization of the subsurface. Dr. Daniels is a member of the American Geophysical Union, the Society of Exploration Geophysicists, and several other professional societies. He holds a B.S. and an M.S. in geology from Michigan State University and a Ph.D. in geophysical engineering from the Colorado School of Mines.

JOHN N. FISCHER is an environmental consultant. His expertise is in groundwater hydrology. His career includes 22 years with the U.S. Geological Survey (USGS) during which time he served as acting associate director, associate chief of the Water Resources Division and the National Mapping Division, and as assistant chief hydrologist for program coordination. In the latter capacity, he was responsible for USGS programs at civilian and DOE radioactive waste disposal sites and at the DOE site at Yucca Mountain. He holds degrees from the U.S. Naval Academy, Michigan State University, and the University of Arizona.

TISSA H. ILLANGASEKARE is the AMAX distinguished chair of environmental sciences and engineering and a professor of civil engineering at the Colorado School of Mines. Until August 1998, he served as a professor of civil and environmental engineering in the Department of Civil Environmental and Architectural Engineering at the University of Colorado, Boulder. His expertise is in numerical modeling of flow and transport in porous and fractured media, multiphase flow modeling, aquifer remediation, and physical modeling of flow and transport in laboratory test tanks. He holds a Ph.D. in civil engineering from Colorado State University. He is also a registered professional engineer and a professional hydrologist.

AARON L. MILLS is a professor of environmental science at the University of Virginia. He has expertise in microbial transformations of organic and inorganic pollutants and bacteria in the subsurface environment. He is a member of the American Geophysical Union, the American Society for Microbiology, and the National Ground Water Association. Dr. Mills holds a B.A. in biology from Ithaca College, and an M.S. in soil science with a minor in microbiology and a Ph.D. in soil science and ecology from Cornell University.

DONALD T. REED is group leader of the Actinide Speciation and Chemistry Group in the Chemical Technology Division at Argonne National Laboratory. He is an expert in radionuclide speciation and migration in subsurface media. He has undertaken a number of basic and applied projects in the fields of actinide speciation, solubility, and subsurface interactions. His most recent research is focused on microbiological-actinide interactions in the subsurface and the application of synchrotron-based methods to the analysis of actinide species in environmental samples. He is a member of the Nuclear Chemistry Division of the American Chemical Society, American Geophysical Union, and the Material Research Society. He holds a Ph.D. in physical chemistry from Ohio State University.

JEROME SACKS is director of the National Institute of Statistical Sciences in Research Triangle Park, North Carolina and a professor at the Institute of Statistics and Decision Sciences at Duke University. His interests include the use of statistical techniques for characterization of subsurface properties. He has served on several National Research Council committees and boards including membership on the NRC Committee on Building an Environmental Management Science Program, which helped the Department of Energy establish its Environmental Management Science Program, the topic of the current study. He has held professorships at the California Institute of Technology, Columbia University, Cornell University, Northwestern University, Rutgers University, University of Illinois, and Duke University. Dr. Sacks has served as program director for statistics and probability at the National Science Foundation. He holds a B.A. and Ph.D. in mathematics from Cornell University.

BRIDGET R. SCANLON is a research scientist in the Bureau of Economic Geology and also teaches courses in the geology and civil engineering departments at the University of Texas at Austin. Her expertise lies in unsaturated zone hydrology, soil physics, environmental tracers, and numerical simulations to quantify subsurface flow in arid regions. She served on the National Research Council Committee on Ward Valley. She has served as a consultant to the Nuclear Waste Technical Review Board. Dr. Scanlon received her Ph.D. in geology at the University of Kentucky.

LEON T. SILVER is a W.M. Keck Foundation professor for resource geology, emeritus, Division of Geological and Planetary Sciences, at the California Institute of Technology. He has expertise in geology, petrology, and geochemistry, with special emphasis on uranium and thorium.

Dr. Silver was a public works officer in the U.S. Naval Civil Engineer Corps from 1945 to 1946, and held several positions at the U.S. Geological Survey before he joined Caltech. He has served on numerous NRC committees, panels and boards, including his past membership on the committee on Building an Environmental Management Science Program. He earned a B.S. in civil engineering from the University of Colorado, an M.S. in geology from the University of New Mexico, and a Ph.D. in geology and geochemistry from the California Institute of Technology. He is a member of the National Academy of Sciences and a past president of the Geological Society of America.

CLAIRE WELTY is associate professor of civil and environmental engineering and associate director and graduate advisor at the School of Environmental Science, Engineering and Policy at Drexel University. She has expertise in groundwater hydrology and contaminant transport. Her current research projects include evaluation of the effects of the interaction between porous medium heterogeneity and fluid density on field-scale dispersion, stochastic analysis of virus transport in aquifers, and tracer tests in fractured sedementary rock. She teaches graduate courses in groundwater hydrology, subsurface contaminant transport, water resources systems analysis, and stochastic subsurface hydrology. Dr. Welty holds a Ph.D. from Massachusetts Institute of Technology.

D

Additional Resources

The following publications provide additional information on the DOE complex and subsurface contamination research and development. The DOE and EM web sites (www.doe.gov; www.em.doe.gov) provide additional information and resources.

1. Closing the Circle on the Splitting of the Atom: The Environmental Legacy of Nuclear Weapons Production in the United States and What the Department of Energy Is Doing About It. Washington, D.C.: U.S. Department of Energy, Office of Environmental Management. 1995.

   The report describes the environmental legacy from the production of nuclear weapons and the cleanup underway by DOE. The report gives a detailed explanation of the nuclear production process and includes information on the extent and types of contaminants produced by each of the steps in the process. The report also describes the types of waste, cleanup actions, and progress made at some DOE sites. The report contains many photographs of the sites and past waste management practices. It also contains a short section on the production of nuclear weapons in other countries, and on environmental contamination in the former Soviet Union.
2. Bioremediation of Metals and Radionuclides: What It Is and How It Works. LBNL-42595. J. McCullough, T.C. Hazen, S.M. Benson, F.B. Metting, and A.C. Palmisano. Lawrence Berkeley National Laboratory. 1995.

   This report explores the possibilities of using bioremediation technology to clean up hazardous metal and radionuclide contaminants found in the DOE complex. Included in the report is an overview of contamination problems at DOE facilities, a summary of some of the most commonly used bioremediation technologies, a discussion of the chemical and physical properties of metals and radionuclides found in contaminant mixtures at DOE sites, an overview of the basic microbial processes that occur in bioremediation, specific in situ bioremediation processes that can be used on these contaminant mixtures, and a hypothetical case study of a composite DOE site with contaminated groundwater.

3. The 1996 Baseline Environmental Management Report. DOE/EM-0290. Washington, D.C.: U.S. Department of Energy, Office of Environmental Management. 1996.

   The report provides an estimate of life-cycle costs and schedules for DOE's environmental cleanup mission. Although the cost and schedule estimates in this report have been superseded by the 1998 *Paths to Closure Report,* the descriptions of waste and contamination problems at DOE sites are still among the most comprehensive published to date.

4. Linking Legacies: Connecting the Cold War Nuclear Weapons Production Processes to Their Environmental Consequences. DOE/EM-0319. Washington, D.C.: U.S. Department of Energy, Office of Environmental Management. 1997.

   The report provides a detailed analysis of the sources of waste and the contamination generated by the production of nuclear weapons, giving specific environmental impacts of particular production activities, in effect "linking" two of DOE's legacies—nuclear weapons manufacturing and environmental management. The report quantifies the current environmental results of past weapons production activities and also contains information on the mission and functions of nuclear weapons facilities, the inventories of waste and materials remaining at these facilities, and the extent and characteristics of contamination in and around these facilities.

5. Accelerating Cleanup: Paths to Closure. DOE/EM-0362. Washington, D.C.: U.S. Department of Energy, Office of Environmental Management. 1998.

   The report outlines DOE's cleanup plans based on site-developed, project-by-project forecasts of the scope, schedule, and costs to complete the more than 300 projects in its cleanup program. The forecasts provide information on technical activities, budgets, worker health and safety, and risk. The report also provides a discussion of the Environmental Management program's decision-making process and the relationship of the "Paths to Closure" plan to that process. Included in the report are summaries of environmental management activities at specific sites, which provide information on the type and extent of the contamination problem, end states, cost and completion dates, remedial actions, and critical closure paths.

6. Groundwater/Vadose Zone Integration Project Specification. DOE/RL-98-48. Draft C. Washington, D.C.: U.S. Department of Energy, Office of Environmental Management. 1998.

   The report describes the Hanford Site's Groundwater/Vadose Zone Integration Project, a science-based strategy established in 1997 to integrate all aspects of the remediation work at Hanford with the ultimate goal of protecting the Columbia River, river-dependent life, and users of the river's resources. Included in the report is a detailed description of the environmental setting of the Hanford Site, its climate and meteorology, geology, hydrology, water quality, and ecology. Also included is a long-range plan for remediation and closure for each of Hanford's main areas (100, 200, and 300 areas). The report appendixes include descriptions of technical elements, the operational history of waste disposal at Hanford, federal and state laws and regulations, current state of technical knowledge, and an applied science and technology plan.

7. Environmental Management Research and Development Program Plan: Solution-Based Investments in Science and Technology. Washington, D.C.: U.S. Department of Energy, Office of Environmental Management. 1998.

   This program plan describes the investments that the Environmental Management (EM) program will make in science and technology to support the DOE cleanup mission. It also describes EM's approach to planning and managing these investments. The plan incorporates what DOE terms "roadmapping" to identify the science and technology areas that promise the greatest return on investment by reducing cleanup project cost, schedule, technical risk, and risk to workers, the public, and the environment. The program plan describes EM's major problem areas, including contaminated environmental soil and groundwater, high-level radioactive waste, spent nuclear fuel, and nuclear materials.

8. Hanford Tank Clean Up: A Guide to Understanding the Technical Issues. R.E. Gephart and R.E. Lundgren. Battelle Press. 1998.

   The report provides a good summary of the basic issues related to high-level radioactive waste that is being stored in 177 underground tanks at the Hanford Site. It provides background information on the history of the site, the production of high-level radioactive waste, the construction of the underground tanks and related facilities, and efforts to manage the waste and associated environmental contamination. The report also details the critical technical issues that need to be addressed for cleanup of the tanks.

9. National Research Council (NRC). Ground Water and Soil Cleanup: Improving Management of Persistent Contaminants. Washington, D.C.: National Academy Press. 1999.

   This report advises DOE on technologies and strategies for cleaning up three types of soil and ground water contaminants: metals, radionuclides, and dense nonaqueous phase liquids. The report describes DOE's program in groundwater and soil remediation, the changing regulatory environment, and technologies being used to remediate each of the contaminant types noted above. Specific advice to DOE suggests ways to set priorities in technology development, to improve the overall technology development program, to overcome barriers to technology deployment, and to address budget limitations.

10. From Cleanup to Stewardship. DOE/EM-0466. Washington, D.C.: U.S. Department of Energy, Office of Environmental Management. 1999.

   This is a companion report to Accelerating Cleanup: Paths to Closure and provides background information on current and planned long-term stewardship activities at DOE sites. The report summarizes what is currently known about end states at DOE sites, and it also provides information on the number and locations of sites that will require continuing management after DOE cleanup is completed. Additionally, the report identifies several issues that will need to be addressed to ensure a successful transition from cleanup to stewardship.

E

Interim Report

NATIONAL RESEARCH COUNCIL
BOARD ON RADIOACTIVE WASTE MANAGEMENT
WATER SCIENCE AND TECHNOLOGY BOARD
2101 Constitution Avenue Washington, D.C. 20418

Executive Office 202/334-3066

December 10, 1998

Mr. Mark Gilbertson
Office of Science and Risk Policy
U.S. Department of Energy
Washington, D.C. 20585

Dear Mr. Gilbertson:

At the request of the Department of Energy's (DOE's) Office of Science and Risk Policy, the National Research Council empaneled a committee[1] to assist the Department in developing a long-range science plan for subsurface contamination research sponsored by the Environmental Management Science Program (EMSP).[2] The committee was asked by DOE to develop an interim report—which is provided in this letter—on the technical content of the EMSP proposal call for fiscal year 1999 (FY99),[3] which DOE intends to focus on subsurface contamination problems. This interim report reflects a consensus of the committee. It has been reviewed in accordance with the procedures of the National Research Council.[4]

The information used to develop this interim report was obtained from several sources. The committee reviewed previous National Research Council reports on the EMSP.[5] The committee also held two information-gathering meetings to familiarize itself with subsurface contamination problems at the five major DOE complex sites: Hanford, Idaho, Oak Ridge, Rocky Flats, and Savannah River. The first meeting, which was held on September 9-10, 1998 in Washington, D.C., provided the committee with an overview of the contamination problems at all five of these sites. The second meeting was held on November 10-12, 1998, in Augusta, Georgia and focused on subsurface contamination problems at the Oak Ridge and Savannah River sites. A third meeting is planned for December 15-17, 1998 to obtain additional information on contamination problems at the Hanford, Idaho, and Nevada Test sites.

The committee also reviewed the portfolio of subsurface contamination-related research projects supported by the EMSP since its inception in 1996.[6] This information included project

---

[1]Committee on Subsurface Contamination at DOE Complex Sites: Research Needs and Opportunities. The roster for this committee is given in Attachment A.

[2]The committee's statement of task is given in Attachment B.

[3]DOE intends to publish the proposal call in the Federal Register in January 1999.

[4]The list of reviewers is given in Attachment C.

[5]Three reports were written by the Committee on Building an Environmental Management Science Program in 1996-97. All three reports are reproduced in the report entitled *Building an Effective Environmental Management Science Program: Final Assessment* (National Research Council, 1997).

[6]Information for this assessment was provided in two Department of Energy reports: U.S. Department of Energy. 1998. *Report to Congress on the U.S. Department of Energy's Environmental Management Science Program.* DOE/EM-0357. Washington, D.C.: DOE Office of Environmental Management; and U.S. Department of Energy. 1998. *Environmental Management Science Program Workshop.* CONF-980736. Washington, D.C.: DOE Office of Environmental Management.

titles, principal investigator names and affiliations, and project abstracts. The purpose of this assessment was to determine the range of research problems being addressed by the program and also to begin the process of identifying potential research gaps. This assessment was conducted by grouping the projects into the following five subsurface contamination problem areas defined by DOE's Subsurface Contaminants Focus Area (SCFA):[7]

- Locate and quantify—Detect and characterize subsurface contamination.
- Contain and stabilize—Eliminate or reduce significantly the migration of contaminants in the subsurface.
- Treat or destroy in situ—Remediate subsurface contamination in place.
- Remove hotspots—Selectively remove highly contaminated zones from the subsurface.
- Validate performance—Confirm the effectiveness of remediation processes or strategies.

These problem areas are being used by the SCFA to organize its subsurface contamination technology development activities. The committee adopted this scheme for organizing its assessment of the EMSP portfolio mainly for convenience, but also because this scheme has the potential to provide a direct linkage between research in the EMSP and technology development in the SCFA. The committee may decide to modify or abandon this scheme as it continues its deliberations.

Given the limited information gathering and deliberations to date, the committee can offer only general advice to DOE on the technical content of the FY99 proposal call. The committee hopes that the following advice will be helpful to the Department:

1. *Focus on basic research.* As noted by previous National Research Council reports (see footnote 5), the purpose of the EMSP is to foster basic research[8] that will contribute to successful completion of DOE's mission to cleanup the environmental contamination across the DOE complex. The committee recommends that DOE articulate clearly the program's focus on basic research—not site-specific remediation problems—in the proposal call.

2. *Focus on subsurface contamination research.* Although the focus of the EMSP is on basic research, as noted above, the objective of this research program is to generate new knowledge to support DOE's mission to remediate its contaminated sites. Some of the Department's most significant contamination problems involve soil and groundwater that contain DNAPLs,[9] metals, and radionuclides. The Department's ability to identify and quantify

---

[7]The Subsurface Contaminants Focus Area is part of the Office of Science and Technology within DOE's Office of Environmental Management, the latter of which has the overall responsibility for cleanup of the weapons complex. Mr. Tom Hicks of the Subsurface Contaminants Focus Area provided the five problem areas in a presentation at the committee's first meeting.

[8]Research that "creates new knowledge; is generic, non-appropriable, and openly available; is often done with no specific application in mind; requires a long-term commitment." (*Allocating Federal Funds for Basic Research*, National Research Council, 1995, p. 6).

[9]DNAPLs, or dense non-aqueous phase liquids, are chlorinated organic solvents such as perchloroethylene and trichloroethylene.

contaminant sources, predict and monitor contaminant fate, and carry out appropriate remediation remains elusive at many sites across the DOE complex. The Department has published several reports that highlight subsurface contamination as a significant long-term problem.[10] Moreover, the EMSP portfolio is well represented by research projects focused on subsurface contamination problems. Thus, DOE's plan to focus the proposal call on subsurface contamination problems seems prudent to the committee in light of the scope of these problems across the complex.

Restricting the proposal call to subsurface contamination problems also seems prudent to the committee in view of the limited funding available to the EMSP. About $10 million will be available to the program in FY99, which will be sufficient to support between 20 and 30 three-year projects.[11] By restricting the proposal call, the Department may be able to approach a "critical mass" of projects in its subsurface contamination research portfolio and thereby make a significant contribution to solving difficult and costly problems at its sites.

3. *Complex-wide focus.* It is apparent to the committee that DOE still faces significant subsurface contamination problems at all five of its major sites. Some problems, like DNAPL and tritium contamination in groundwater, are common to all five sites, whereas other problems, such as mercury contamination in soil, appear to be less common across the complex. Moreover, all five sites have different geological, hydrological, and climatic conditions and, thus, are in some senses unique. If the EMSP is to make a significant long-term contribution to the Department's mission to cleanup *all* of its sites, the proposal call should encourage the submission of research ideas that address significant subsurface contamination problems across the complex. That is, the proposal call should encourage the submission of proposals that tackle significant science problems that are relevant to any DOE site.

A proposal call with a complex-wide focus would have at least one practical benefit for the EMSP—namely, it likely would increase the quality of the proposal pool. A complex-wide call likely would generate a better selection of proposals from researchers across the nation. The Department could then use its merit and relevance review processes to select for funding those projects that are likely to have highly significant impacts on both science and the cleanup mission as a whole. A more restricted proposal call likely would attract only proposals from researchers who happened to be acquainted with problems at the sites covered in the call. A proposal call with a complex-wide focus would increase competition among research ideas and thereby increase the overall quality of the EMSP research portfolio.

Although the committee recommends a complex-wide focus for the proposal call, it also believes that researchers should be encouraged to demonstrate a linkage between their research projects and significant contamination problems at DOE sites. Researchers could

---

[10]See, for example, U.S. Department of Energy. 1997. *Linking Legacies: Connecting the Cold War Nuclear Weapons Production Processes to Their Environmental Consequences.* DOE/EM-0319. Washington, D.C.: DOE Office of Environmental Management; and U.S. Department of Energy. 1998. *Accelerating Cleanup: Paths to Closure.* DOE/EM-0362. Washington, D.C.: DOE Office of Environmental Management.

[11]Information received from Mark Gilbertson, Director of the Office of Science and Risk Policy, at the committee's second meeting.

establish this linkage in a variety of ways—for example, by elucidating the scientific problems to be addressed by the proposed research and explaining how the solution of these problems could improve remediation capabilities. Of course, given the nature of basic research, there will not always be a clear pathway between research results and application to site remediation. Nevertheless, the committee believes that this linkage exercise will help researchers focus their proposals on those key scientific problems that have significant implications for site remediation and, moreover, that the linkage information provided in the proposals will help the Department assess project relevance.

4. *Science problems in the proposal call.* Although the committee concurs with DOE's plan to focus the proposal call on subsurface contamination problems, as noted previously, the committee is not yet ready to make specific recommendations on a science plan for subsurface contamination research—that plan will be the subject of the committee's final report. Therefore, the committee believes that the call should be written to encourage the submission of new and innovative basic research ideas that address science problems relevant to all five of the subsurface contamination problem areas described above.

In its preliminary assessment of the EMSP portfolio, the committee has observed that there are relatively few basic research projects in the *validate performance* and possibly the *remove hotspots* problem areas, although the committee's assessment of the latter category is continuing. In the context of the EMSP, *validate performance* concerns the ability to confirm the performance or behavior of a physical, chemical, or biological process or a technology at a contaminated site. Basic science can contribute to performance validation through the investigation and development of new or improved tools and methodologies for confirming behavior or performance in the field. There are a number of underlying theoretical and experimental issues of interest—for example, understanding the pre-remediation conditions at a contaminated site and the fundamental hydrogeological, chemical, and biological controls on site or contaminant behavior, how these change during site remediation, and which tests or measurements are sensitive to the behaviors of concern.

The inability to confirm such behavior or performance at a contaminated site is one of the primary reasons for the Department's difficulty in prescribing appropriate and cost-effective remediation and monitoring strategies.[12] Moreover, once a remediation action is underway, the Department often lacks methods to measure and confirm the efficacy of the approach. Deployment of new remediation technologies may depend to a great extent on the Department's ability to validate their effectiveness—and provide evidence of remediation efficacy to regulators and other stakeholders.

The committee views the basic science issues underlying the validate performance problem area as a research opportunity for the EMSP. This problem area is under-represented in the current EMSP portfolio, and new knowledge obtained through the program could lead to significant improvements in remediation capabilities.

---

[12]The idea that lack of process validation can limit technology application also is discussed in the National Research Council report entitled *Innovations in Ground Water and Soil Cleanup* (National Research Council, 1997).

Although the committee recommends that the call focus on new and innovative research proposals on the entire spectrum of subsurface contamination problems, the committee suggests that the call also indicate DOE's receptiveness to the submission of new research ideas that address the basic science aspects of performance validation.

Sincerely,

Jane C.S. Long, Chair
James K. Mitchell, Vice-Chair

Attachment A: Roster of Committee Members
Attachment B: Statement of Task
Attachment C: List of Reviewers

ATTACHMENT A
COMMITTEE ROSTER

COMMITTEE ON SUBSURFACE CONTAMINATION AT DOE COMPLEX SITES: RESEARCH NEEDS AND OPPORTUNITIES

JANE C.S. LONG, CHAIR, Mackay School of Mines, University of Nevada, Reno
JAMES K. MITCHELL, VICE CHAIR, Virginia Polytechnic Institute and State University, Blacksburg
RANDALL J. CHARBENEAU, University of Texas, Austin
JEFFREY J. DANIELS, The Ohio State University, Columbus
JACK N. FISCHER, Hydrologic Consultant, Oakton, Virginia
TISSA H. ILLANGASEKARE, Colorado School of Mines, Golden
AARON L. MILLS, University of Virginia, Charlottesville
DONALD T. REED, Argonne National Laboratory, Illinois
JEROME SACKS, National Institute for Statistical Sciences, Research Triangle Park, North Carolina
BRIDGET R. SCANLON, Bureau of Economic Geology, University of Texas, Austin
LEON T. SILVER, California Institute of Technology, Pasadena
CLAIRE WELTY, Drexel University, Philadelphia, Pennsylvania

Staff

KEVIN D. CROWLEY, Study Director and Director, Board on Radioactive Waste Management
STEPHEN D. PARKER, Director, Water Science and Technology Board
SUSAN B. MOCKLER, Research Associate, Board on Radioactive Waste Management
PATRICIA A. JONES, Senior Project Assistant, Board on Radioactive Waste Management

ATTACHMENT B
STATEMENT OF TASK

COMMITTEE ON SUBSURFACE CONTAMINATION AT DOE COMPLEX SITES: RESEARCH NEEDS AND OPPORTUNITIES

The objective of this study is to develop a science plan for subsurface contamination research sponsored by DOE's EM Science Program. This science plan will describe the significant subsurface contamination problems at DOE sites that cannot be addressed with current technologies, identify the knowledge gaps relevant to these problems, and develop a research plan to fill these gaps. This plan will take account of research being sponsored by other federal and state agencies and will identify those areas of research where the EM Science Program can make significant contributions to addressing DOE's problems and adding to scientific knowledge generally.

ATTACHMENT C
LIST OF REVIEWERS

This letter report has been reviewed in draft form by individuals chosen for their diverse perspectives and technical expertise, in accordance with procedures approved by the NRC's Report Review Committee. The purpose of this independent review is to provide candid and critical comments that will assist the institution in making the published report as sound as possible and to ensure that the report meets institutional standards for objectivity, evidence, and responsiveness to the study charge. The review comments and draft manuscript remain confidential to protect the integrity of the deliberative process. We wish to thank the following individuals for their participation in the review of this report:

John F. Ahearne, Sigma Xi, The Scientific Research Society and Duke University, Research Triangle Park, North Carolina
Richelle Allen-King, Washington State University, Pullman
George M. Hornberger, University of Virginia, Charlottesville
Richard G. Luthy, Carnegie Mellon University, Pittsburgh, Pennsylvania
Norine E. Noonan, U.S. Environmental Protection Agency, Washington, D.C.

While the individuals listed above have provided constructive comments and suggestions, it must be emphasized that responsibility for the final content of this report rests entirely with the authoring committee and the institution.

F

Acronym List

DNAPL	Dense non-aqueous phase liquid
DOD	U.S. Department of Defense
DOE	U.S. Department of Energy
EM	Environmental Management
EPA	U.S. Environmental Protection Agency
NAPL	Non-aqueous phase liquid
NRC	National Research Council
NSF	National Science Foundation
USGS	U.S. Geological Survey